Applied Condition Monitoring 23

The book series Applied Condition Monitoring publishes the latest research and developments in the field of condition monitoring, with a special focus on industrial applications. It covers both theoretical and experimental approaches, as well as a range of monitoring conditioning techniques and new trends and challenges in the field. Topics of interest include, but are not limited to: vibration measurement and analysis; infrared thermography; oil analysis and tribology; acoustic emissions and ultrasonics; and motor current analysis. Books published in the series deal with root cause analysis, failure and degradation scenarios, proactive and predictive techniques, and many other aspects related to condition monitoring. Applications concern different industrial sectors: automotive engineering, power engineering, civil engineering, geoengineering, bioengineering, etc. The series publishes monographs, edited books, and selected conference proceedings, as well as textbooks for advanced students.

** Indexing: Indexed by SCOPUS, WTI Frankfurt eG, SCImago

Ahmed Hammami · Abdel Kader Mahmoud · Ahmed Mohamed Yahya · Philippus Stephanus Heyns · Majdi Yangui · Mohamed Haddar
Editors

Modelling and Simulation of Complex Systems for Sustainable Energy Efficiency II

Selected Contributions to the 2nd International Workshop on Modelling and Simulation of Complex Systems for Sustainable Energy Efficiency, MOSCOSSEE'2024, November 2–3, 2024, Matmata, Gabes, Tunisia

Springer

Editors
Ahmed Hammami
National School of Engineers of Gabes
Gabes, Tunisia

Ahmed Mohamed Yahya
URA3E - EX CRAER
University of Nouakchott
Nouakchott, Mauritania

Majdi Yangui
National School of Engineers of Sfax
Sfax, Tunisia

Abdel Kader Mahmoud
URA3E - EX CRAER
University of Nouakchott
Nouakchott, Mauritania

Philippus Stephanus Heyns
Department of Mechanical and Aeronautical Engineering
University of Pretoria
Pretoria, South Africa

Mohamed Haddar
National School of Engineers of Sfax
Sfax, Tunisia

ISSN 2363-698X ISSN 2363-6998 (electronic)
Applied Condition Monitoring
ISBN 978-3-032-00929-6 ISBN 978-3-032-00930-2 (eBook)
https://doi.org/10.1007/978-3-032-00930-2

This Springer imprint is published by the registered company Springer Nature Switzerland AG
The registered company address is: Gewerbestrasse 11, 6330 Cham, Switzerland

Preface

This volume of Applied Condition Monitoring gathers selected papers presented at the second "International Workshop on MOdelling and Simulation of COmplex Systems for Sustainable Energy Efficiency" MOSCOSSEE'2024, organized by the LAboratory of Mechanics, Modelling and Production (LA2MP) from University of Sfax, Tunisia, the Department of Mechanical Engineering at the National School of Engineers of Gabes from the University of Gabes, Tunisia, and the and the Applied Research Unit for Renewable Energy, Water and Environment from the University of Nouakchott, Mauritania, and held in Matmata, Tunisia, on November 02–03, 2024. This workshop is in the framework of the Tunisian-Mauritanian project in "23MT1003" entitled "Design, modelling, simulation and remote monitoring of hybrid renewable energy systems for improved sustainable energy efficiency".

This book is also a tribute to the memory of Professor Fakher Chaari, editor of the series "Applied Condition Monitoring" and "Lecture Notes in Mechanical Engineering", editor of the previous proceedings publication "Modelling and Simulation of Complex Systems for Sustainable Energy Efficiency", and chairman and main organizer of the first edition of this workshop.

During MOSCOSSEE'2024 Workshop, high-level contributions in the field of complex systems for sustainable energy efficiency were presented. Three plenary sessions were held by eminent scientists who kindly agreed to share their knowledge in the field of the workshop.

This book gathers 20 chapters selected out of all the papers presented at the workshop after a rigorous peer-review process. 25 submissions were received from Tunisia, Mauritania, South Africa, Iraq, Spain, Portugal and France. Each paper was reviewed by at least two independent reviewers. Eight papers were reviewed by three independent reviewers, and one paper was reviewed by four independent reviewers. The reviewers had to evaluate each paper by answering questions about the originality of the paper, the significance of the results presented, the clarity and presentation of the chapters and the relevance to this conference. The reviewers came from ten countries in Africa, Asia and Europe. iThenticate was used for plagiarism check. After this process, authors were asked to revise their paper according to the reviewers' comments. Eventually, 20 papers written by 42 authors from 7 countries were accepted for inclusion in this book

During the two days of the workshop, oral communications discussed several topics such as:

Hybrid Energy Systems
Sustainable and Renewable Energies
Fluid-Structure Interaction
Mechatronics
Dynamics and Engines Control
Structural Vibration and Acoustics
PDEs of Multiphysics Dynamic Systems

Smart Materials and Meta-composites
Fault Diagnosis and Prognosis
Robust Design

The editors would like to thank all the participants from Tunisia, Mauritania, South Africa, Spain and France, as well as the reviewers of the book chapters. We acknowledge the financial support of the Ministry of Higher Education and Scientific Research of Tunisia and the Ministry of Higher Education of Mauritania, the National Agency for Scientific Research and Innovation (ANRSI) of Mauritania in the framework of the Tunisian-Mauritanian project in "23MT1003". We would also like to thank Springer for their support of the MOSCOSSEE'2024 workshop.

April 2024

Ahmed Hammami
Abdel Kader Mahmoud
Ahmed Mohamed Yahya
Phillipus Stephanus Heyns
Majdi Yangui
Mohamed Haddar

Organization

General Chairs

Ahmed Hammami	National School of Engineers of Gabes, Tunisia
Abdel Kader Mahmoud	University of Nouakchott, Mauritania

Organizing Committee Chairs

Atef Hmida	National School of Engineers of Gafsa, Tunisia
Ahmed Mohamed Yahya	University of Nouakchott, Mauritania

Organizing Committee

Maroua Haddar	National School of Engineers of Sousse, Tunisia
Chaima Hammami	Higher Institute of Applied Sciences and Technology of Sousse, Tunisia
Marwa Hammami	National School of Engineers of Carthage, Tunisia
Mohamed El Khayar Mahdi	University of Nouakchott, Mauritania
Ayoub Mbarek	National School of Engineers of Bizerte, Tunisia
Majdi Yangui	National School of Engineers of Sfax, Tunisia
Olfa Kessentini	Higher Institute of Industrial Systems of Gabes, Tunisia

Scientific Committee

Abbes Mohamed Slim	University of Sfax, Tunisia
Abdennadher Moez	University of Sfax, Tunisia
ABID Said	University of Sfax, Tunisia
Aifaoui Nizar	University of Sousse, Tunisia
Akrout Ali	University of Sfax, Tunisia
Aloui Abdessattar	University of Gabes, Tunisia
Azari Zitoune	University of Lorraine, France
Ayadi Omar	University of Sfax, Tunisia
Bahloul Riadh	University of Sousse, Tunisia
Barkallah Maher	University of Sfax, Tunisia

Battarra Mattia	University of Ferrara, Italia
Benamar Mounir	University of Paris 13, France
Bentahar Mabrouk	University of Technology of Compiègne, France
Benjdidia Mounir	University of Sfax, Tunisia
Bensalah Nizar	University of Tunis, Tunisia
Ben Souf Mohamed Amine	University of Sfax, Tunisia
Beyaoui Moez	University of Sfax, Tunisia
Bouabidi Abdallah	University of Gabes, Tunisia
Bouaziz Slim	University of Sfax, Tunisia
Bouaziz Zoubeir	University of Sfax, Tunisia
Bougacha Anas	University of Sfax, Tunisia
Bouraoui Chokri	University of Sousse, Tunisia
Bouraoui Tarek	University of Monastir, Tunisia
Chaari Riadh	University of Sfax, Tunisia
Chiementin Xavier	University of Reims, France
Chouchane Mnaouar	University of Monastir, Tunisia
Chrigui Mouldi	University of Gabes, Tunisia
Dammak Fakhreddine	University of Sfax, Tunisia
Debut Vincent	Nova University Lisbon, Portugal
De-Juan Ana	University of Cantabria, Spain
Dewangan Palash	University of Huddersfield, UK
Diez Alberto	University of Cantabria, Spain
Djemal Fathi	University of Gabes, Tunisia
Driss Zied	University of Sfax, Tunisia
Dourado Nuno	University of Minho, Portugal
El Aoud Sami	University of Sfax, Tunisia
Fakhfakh Tahar	University of Sfax, Tunisia
Fedala Semcheddine	FEDALA Semcheddine
Feki Nabih	University of Sousse, Tunisia
Felkaoui Ahmed	FEDALA Semcheddine
Fernandez Alfonso	University of Cantabria, Spain
Garcia Pablo	University of Cantabria, Spain
Ghorbel Elhem	Cergy-Pontoise University, France
Haddar Mohamed	University of Sfax, Tunisia
Hammami Lotfi	University of Sfax, Tunisia
Hammami Ahmed	University of Gabes, Tunisia
Hentati Taissir	University of Sfax, Tunisia
Hentati Hamdi	University of Sousse, Tunisia
Heyns Stephan	University of Pretoria, South Africa
Jarraya Abdessalem	University of Jeddah, KSA
Karra Chafik	University of Sfax, Tunisia
Kchaou Mounir	University of Sfax, Tunisia

Khabou Mohamed	University of Sfax, Tunisia
Krot Pavlo	Wroclaw University of Science and Technology, Poland
Louati Jamel	University of Sfax, Tunisia
Mahmoud Abdel Kader	University of Nouakchott, Mauritania
Mucchi Emiliano	University of Ferrara, Italia
Nouri Ahmed Said	University of Sfax, Tunisia
Parey Anand	University of Indore, India
Saka Abdelmajid	Sidi Mohamed Ben Abdellah University, Morocco
Schmidt Stephan	University of Pretoria, South Africa
Soula Mohamed	University of Tunis, Tunisia
Taktak Mohamed	University of Sfax, Tunisia
Trabelsi Hassen	University of Sfax, Tunisia
Trabelsi Hassen	University of Gabes, Tunisia
Viadero Fernando	University of Cantabria, Spain
Walha Lassaad	University of Sfax, Tunisia
Wilke Daniel Nicolas	University of Pretoria, South Africa
Yahya Ahmed Mohamed	University of Nouakchott, Mauritania
Ziani Ridha	University of Setif, Algeria
Zouari Bassem	University of Sfax, Tunisia

Contents

Modelling and Control of Renewable Energy Systems

Using MFC Strategy to Maximize Power Capture for Variable Speed Wind Energy Conversion Systems ... 3
Maroua Haddar, Ahmed Hammami, and Mohamed Haddar

Nature-Inspired Metaheuristic Approaches for Optimizing P&O Algorithm for MPPT Control in Photovoltaic Systems ... 13
Youssef Dhieb, Walid Ouled Amor, and Moez Ghariani

Application of Intelligent Controller to Mechanical Speed Variation in the Case of Wind Turbine ... 26
Rania Maktouf, Maroua Haddar, and Ahmed Hammami

Effect of Cylinder Count on the Torques Generated by Internal Combustion Engines ... 36
Zied Abdelmoula, Ahmed Hammami, and Mohamed Haddar

Modelling of an 18-Cylinder Diesel Engine: Case Study of the 180 MW Power Plant ... 54
Amira Hattay, Abdel Kader Mahmoud, Ahmed Hammami, Mohamed Haddar, and Ahmed Mohamed Yahya

Determination of the Optimum Design of One-Dimensional Wind Turbine Under the Weather Conditions of Mosul City ... 64
Yaser S. H. Mahmood and Mohamed Taktak

Contribution of Profile Shifting on Wind Turbine Gearbox Efficiency ... 80
Abdesselem Toumi, Alberto Diez Ibarbia, Ahmed Hammami, Alfonso Fernandez Del Rincon, Fernando Viadero Rueda, and Mohamed Haddar

Dynamic Analysis of Pre-Twisted Blade Rotor ... 90
Majdi Yangui, Houssem Cheniour, Slim Bouaziz, and Mohamed Haddar

A Comparison of Two Configuration of Solar Water Heating Systems ... 98
Cheikh Brahim Mohameden, Maroua Haddar, Ahmed Hammami, Abdel Kader Mahmoud, Mohamed Haddar, and Ahmed Mohamed Yahya

Influence of Tank Top-Bottom Shape of a Solar Water Storage Tank on the Thermal Stratification 109
Ahmed Samet

The Use of Discrete Random Separation DRS for Fault Detection in a Wind Turbine Gearbox 118
Ayoub Mbarek, Ahmed Hammami, and Fakher Chaari

Contribution to the Study of Dynamic Behavior of Wind Turbine Gearbox 126
Ahmed Hammami, Ayoub Mbarek, Fakher Chaari, Abdel Kader Mahmoud, Mohamed Haddar, and Ahmed Mohamed Yahya

Dynamics of mechanical systems

Investigation of the Crack Impact on a Polymer Spur Gear Performance 137
Abderrahim Baccar, Ahmed Hammami, Nuno Dourado, Ali Akrout, and Fakher Chaari

Investigation of the Influence of Viscoelastic Layer Thickness and Stacking Configurations on the Dynamic Behavior of Flax/PLA Biocomposites 146
Hajer Daoud, Firas Meddeb, Abderrahim ElMahi, Jean-Luc Rebiere, Mohamed Amine Ben Souf, and Mohamed Haddar

Enhancing Lifespan Predictions for Shaft Bearings Using Feed-Forward Neural Networks 154
Rawnak Omar, Chaima Chikhaoui, Fathi Djemal, and Hassen Trabelsi

Impact of Design Parameters on the Dynamic Behavior of Composite Shafts ... 168
Salwa Benali, Taissir Hentati, and Slim Bouaziz

Integration of Artificial Intelligence: An Innovative Synergy for Vibration Reduction of Car Suspension System 178
Fatma Hmida, Hassen Trabelsi, Fathi Djemal, and Rawnak Omar

Modelling of a Moving Load on a Polymer Spur Gear Tooth Profile Using the FEM 190
Abderrahim Baccar, Ahmed Hammami, Nuno Dourado, Ali Akrout, and Fakher Chaari

A Generalised Stiffness Model for Incorporating Multiple Spur Gear Tooth Faults - Part 1: Stiffness Modelling 200
Luke van Eyk, P. Stephan Heyns, and Stephan Schmidt

A Generalised Stiffness Model for Incorporating Multiple Spur Gear Tooth Faults - Part 2: Fault Modelling 216
Luke van Eyk, P. Stephan Heyns, and Stephan Schmidt

Author Index 229

Modelling and Control of Renewable Energy Systems

Using MFC Strategy to Maximize Power Capture for Variable Speed Wind Energy Conversion Systems

Maroua Haddar[1,2(✉)], Ahmed Hammami[1,3], and Mohamed Haddar[1]

[1] Mechanics, Modeling and Production Laboratory (LA2MP), Mechanic Department, National Engineering School of Sfax (ENIS), 1173-3038 Sfax, BP, Tunisia
maroua.haddar@enis.tn

[2] National School of Engineers of Sousse, University of Sousse, 4003 Sousse, Tunisia

[3] National School of Engineers of Gabes, University of Gabes, 6029 Gabes, Tunisia

Abstract. This chapter focuses on power generation control within variable-speed wind turbines, specifically addressing the distinct operational regions determined by the turbine's tip speed ratio. The research aims to maximize power extraction during low wind speed conditions through the implementation of an innovative model-free control strategy. By leveraging an ultra-local model combined with an intelligent proportional integrated derivative controller, this approach delivers significant advantages, including robust resilience against unmodeled dynamics and simplified parameter tuning processes. The study employs an algebraic estimator for ultra-local dynamics to overcome challenges associated with limited accessibility to precise wind turbine models, facilitating straightforward implementation of the control law. The effectiveness of the proposed approach in achieving accurate power tracking is demonstrated under variable-speed wind energy conditions. Performance analysis reveals significant optimization potential through parameter tuning. When configured with a sliding window of L = 0.05 s, the system achieves optimal results, generating 26.95 MWh (representing 99.81% of reference energy) with peak efficiency of 99.81% and minimal tracking error of 100.30 kW. Comprehensive evaluation identifies L = 0.05 s as the optimal parameter configuration, effectively balancing energy production capabilities with tracking precision requirements. This optimization represents a significant advancement in wind turbine control technology, enabling more efficient renewable energy harvesting across variable operating conditions.

Keywords: Model Free Control · intelligent technique · Online estimator · renewable energy

1 Introduction

Renewable energy technologies have emerged as critical solutions to global energy challenges, with wind energy representing a pivotal domain of sustainable power generation. The conversion of wind's kinetic energy into electrical power through wind turbine systems presents a sophisticated engineering challenge that demands advanced control methodologies to optimize performance and efficiency.

A. Hammami et al. (Eds.): MOSCOSSEE 2024, ACM 23, pp. 3–12, 2025.
https://doi.org/10.1007/978-3-032-00930-2_1

Among the diverse wind turbine architectures, variable-speed variable-pitch configurations represent a sophisticated technological approach to mitigating the inherent variability of wind resource dynamics (Roga et al. 2022). These systems integrate complex control mechanisms that dynamically modulate rotor speed and blade pitch angles, enabling comprehensive power extraction across disparate wind speed regimes. However, the implementation of such control strategies is substantially complicated by the nonlinear and stochastic nature of wind energy systems (Song et al. 2018).

Conventional control paradigms, exemplified by proportional-integral-derivative (PID) control, have demonstrated significant limitations in addressing the multifaceted challenges presented by wind turbine dynamics (Jeon, 2024). Researchers have consequently pivoted towards exploring more advanced control strategies, encompassing both model-based and model-free approaches. Model-based methodologies, such as backstepping and adaptive control, necessitate comprehensive system modeling but are fundamentally constrained by computational complexity and sensitivity to parametric uncertainties (Sedaghat et al. 2017).

Model-free control (MFC) emerges as a particularly promising alternative, circumventing the traditional requirements of precise system modeling. By leveraging data-driven learning and adaptive techniques, MFC enables real-time system identification and control optimization. This approach is especially pertinent in wind energy systems, where atmospheric turbulence, topographical variations, and stochastic wind characteristics render deterministic modeling inherently challenging. The model-free control strategy delivers superior power tracking with exceptional robustness and simplified implementation, eliminating model dependencies while maintaining outstanding real-time performance surpassing the industry-standard $K\omega^2$ approach which offers only moderate resilience despite minimal complexity (Bouchemal et al. 2024). While extremum-seeking control matches MFC's robustness without model requirements, it sacrifices tracking speed despite potential steady-state optimization advantages. Model predictive and nonlinear backstepping controls achieve remarkable precision but demand intricate model-dependent architectures with reduced real-time viability. Fuzzy logic control parallels intelligent PID's tracking excellence and robustness with reasonable complexity, though its intuition-based calibration contrasts with MFC's straightforward design (Bustan and Moodi, 2020). Neural network adaptive control demonstrates impressive tracking and adaptability capabilities but imposes substantial computational burdens (Jafarnejadsani et al., 2013), positioning MFC as the preferred solution for constrained systems. Ultimately, MFC strikes an optimal balance between performance and practicality, competing with sophisticated methodologies while distinguishing itself through operational simplicity and real-time efficiency.

The present research introduces a novel model-free control strategy specifically designed for variable-speed wind turbines. The proposed approach integrates ultra-local modeling techniques with an intelligent PID framework, aiming to simultaneously maximize power output and ensure robust operational stability.

The subsequent chapters are organized as follows: Sect. 2 presents a comprehensive exposition of the proposed model-free control strategy. The third section details the wind turbine mathematical model and simulation experimental setup. Section 4 provides a rigorous analysis of simulation results, empirically validating the proposed controller's performance. Section 5 concludes the research by synthesizing key findings and delineating potential trajectories for future investigative endeavors.

2 Model Free Control

Model-Free Control (MFC) represents an innovative approach to dynamic systems control, emerging as a sophisticated alternative to traditional control methodologies. The primary objective of this intelligent control technique is the real-time rejection of unmodeled perturbations, addressing a critical challenge in complex system management (Haddar et al. 2023).

The fundamental innovation of MFC lies in its utilization of ultra-local models, which distinguishes it from conventional control strategies. The method introduces an intelligent Proportional-Integral-Derivative (i-PID) controller characterized by an adaptive tuning parameter. A key distinguishing feature is the incorporation of an additional term F, which fundamentally alters the control formulation compared to classical PID controllers, as illustrated in the referenced Fig. 1. The proposed technique of control offers great conceptual simplicity, as it does not require precise knowledge of the system's mathematical model, relying only on real-time estimations from inputs and outputs to track the reference.

The methodological advancement of MFC lies in its ability to:

- Dynamically adapt to system uncertainties.
- Eliminate the need for precise a priori system modeling.
- Provide real-time perturbation rejection.
- Enhance control system flexibility across diverse operational contexts.

The mathematical model is reduced to a simple ultra-local model:

$$y^{(n)}(t) = F(t) + \alpha u(t) \tag{1}$$

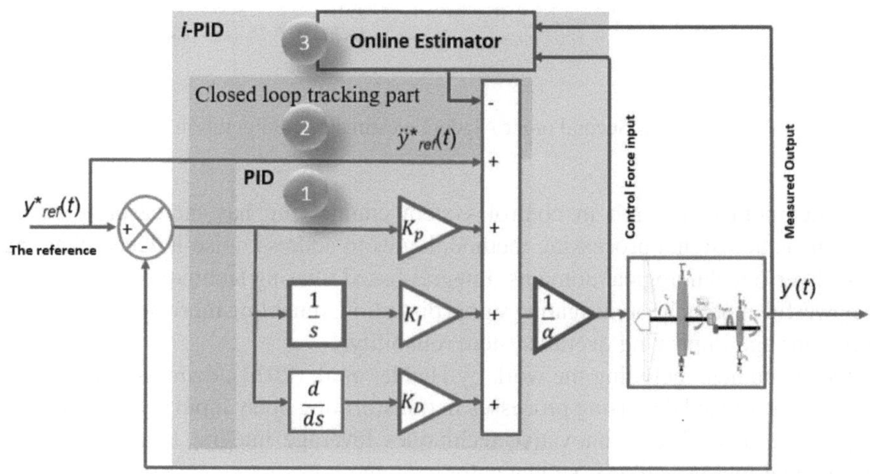

Fig. 1. Block Diagram of Model Free Control

where y is a single output variable and u a single control variable.

n is an integer that is selected by the opperatot (1 or 2).

Empirical studies have demonstrated the potential superiority of second-order models in MFC implementation (n = 2), particularly for systems exhibiting complex dynamic characteristics. These models offer enhanced representation of system behaviors, effectively capturing nuanced phenomena such as acceleration and higher-order dynamic effects. The second-order approach provides a more comprehensive framework for modeling intricate system interactions, thereby improving control precision and adaptive capabilities by using a second order algebraic estimator. Figure 2 details the process for getting the proposed algebraic estimator for the ultra-local model. This approach represents a significant advancement in signal processing technology, offering a robust solution to the persistent challenge of managing uncertain and noisy input signals in complex engineering environments.

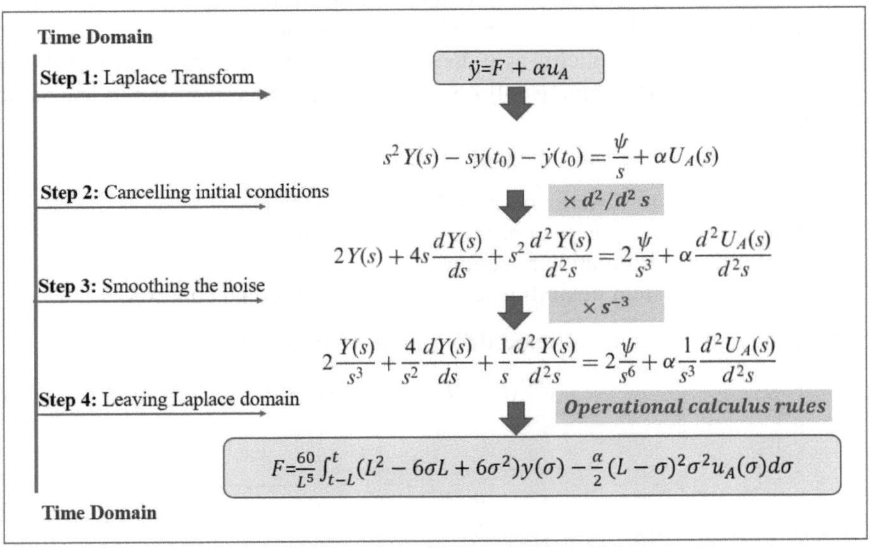

Fig. 2. Second order Algebraic estimator configuration

Contemporary research in control system engineering has increasingly focused on sophisticated signal processing methodologies to address noise-induced challenges across diverse technological domains. Integral-based filtering techniques have emerged as a powerful approach for mitigating signal distortions, enabling more precise derivative estimations and improving overall system reliability.

Recent studies, including the work by Haddar et al. (2023), demonstrate the effectiveness of advanced denoising processes in transforming noisy input signals into clean, interpretable data. These innovative techniques leverage integral transformations as sophisticated low-pass filters, strategically reducing high-frequency noise components while preserving critical signal characteristics.

The algebraic estimator's implementation requires real-time measurement acquisition, continuously monitoring system inputs and outputs. Advanced numerical differentiation methods extract derivative estimates from noise-contaminated data without necessitating comprehensive system modeling. Robustness enhancement incorporates sophisticated filtering techniques to effectively attenuate measurement noise interference. The theoretical continuous-time framework undergoes discretization for digital controller deployment, enabling seamless integration with model-free control architectures including intelligent PID implementations.

3 Wind Turbine Model

Accurately modeling the various interactions of a variable-speed wind turbine are challenged. In fact, there is an inherent complexity of this non-linear model. For controller design purposes, it is common to use a simplified model (See Fig. 3). To overcome the inability to identify the system accurately, this simplification is necessary. As per the model-free control principle discussed earlier, the ultra-local model mentioned in Sect. 2 can be used to approximate the nonlinear dynamics of power as following:

$$\ddot{P}_g(t) = F(t) + \alpha T_g(t) \tag{2}$$

where: 2

- P_g is the generated power
- T_g is the generator torque
- F includes all the unknown dynamics variations
- α is a scaling parameter chosen by the operator.

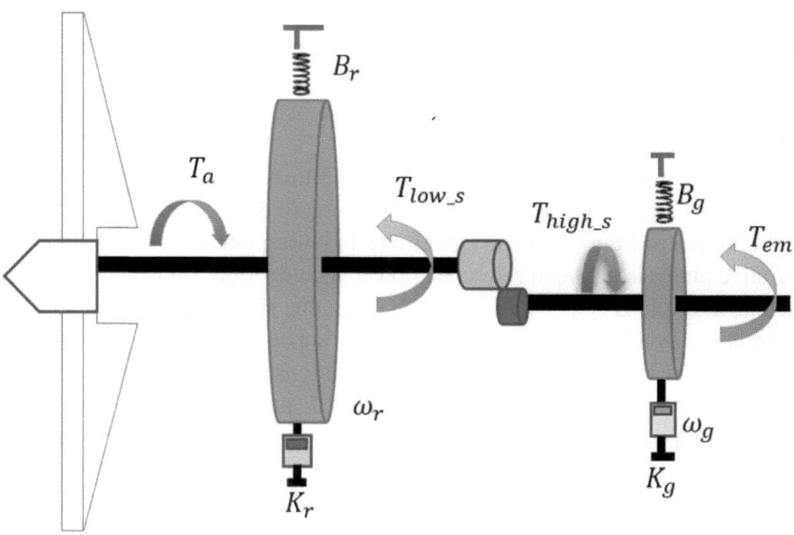

Fig. 3. Simplified model of wind turbine

The algebraic estimator of F is designed as follows according to steps detailed in Fig. 2 in Sect. 2:

$$\widehat{F} = \frac{60}{L^5} \int_{t-L}^{t} \left(L^2 - 6\sigma L + 6\sigma^2\right) P_g(\sigma) - \frac{\alpha}{2}(L-\sigma)^2 \sigma^2 T_g(t) d\sigma \tag{3}$$

where L is a scaling parameter that is known by the sliding window.

The optimal window of integration is determined by its minimal length, L, which optimizes both accuracy and computational efficiency. This length is dependent on the sampling time interval. Building upon this foundation, we can proceed with the implementation of the i-PID (intelligent Proportional–Derivative) controller.

$$T_g(t) = \frac{-\widehat{F}(t) + K_p e(t) + K_i \int e(t) + K_d \dot{e}(t) + \ddot{P}_g^{ref}}{\alpha} \tag{4}$$

where K_p, K_i and K_d are the conventional tuning gains of PID controller

4 Simulation Results

This study models a wind turbine system to analyze power regulation under varying wind conditions. The aerodynamic torque is computed using a power coefficient Cp based on the tip-speed ratio, while the rotor dynamics incorporate external damping and generator torque constraints. The turbine has a rotor radius of R = 70 m, with air density ρ = 1.225 kg/m^3. The total system inertia is Jr = 4.4532 × 10^5 kg·m^2, and the gear ratio is ng = 85. The power regulation strategy considers a power reference ratio np = 0.9, with generator torque saturation at Tgmax = 4300 × 10^3 N·m. The wind speed profile is generated with a mean velocity of v_{mean} = 9 m/s and turbulence intensity of 15%. The performance of the system is analyzed using different size of sliding windows L ranging from 0.05 to 0.5, impacting the response of the generated power to the reference. The model is simulated over 600 s with a sampling time of 0.01 s, capturing key performance metrics like efficiency and power tracking error.

The graphical representations showcase wind turbine control system performance across various sliding window parameters (L). Figure 4 displays temporal wind velocity fluctuations resulting from atmospheric turbulence phenomena. Figure 5 presents a comparative analysis between reference power specifications and actual power generation across multiple L configurations, illustrating this scaling coefficient's influence on power tracking precision. Figure 6 provides an enhanced magnification of the aforementioned comparison, revealing nuanced variations in tracking performance otherwise indiscernible at standard resolution. Figure 7 quantifies the tracking differential, illustrating numerical deviations between target and produced power outputs, while the concluding visualization offers a magnified perspective of these error measurements to facilitate detailed examination.

Decreasing the L size demonstrates progressive enhancement in wind turbine power control performance metrics. At L = 0.25 s, the system generates 26.8 MWh, achieving

99.26% of reference energy with corresponding 99.26% efficiency and maximum tracking deviation of 120.50 kW. When L is adjusted to 0.15s, energy production improves marginally to 26.9 MWh, representing 99.63% of reference energy, accompanied by efficiency improvements to 99.63% and reduced maximum tracking error of 110.20 kW. Further optimization to L = 0.05s yields optimal results with 26.95 MWh generation (99.81% of reference), peak efficiency of 99.81%, and minimal tracking error of 100.3 kW. Analysis identifies L = 0.05s as the optimal parameter configuration, effectively balancing energy production and tracking precision. However, consideration must be given to L's sensitivity to noise perturbations, an aspect not evaluated in this investigation that may impact operational performance in practical implementations.

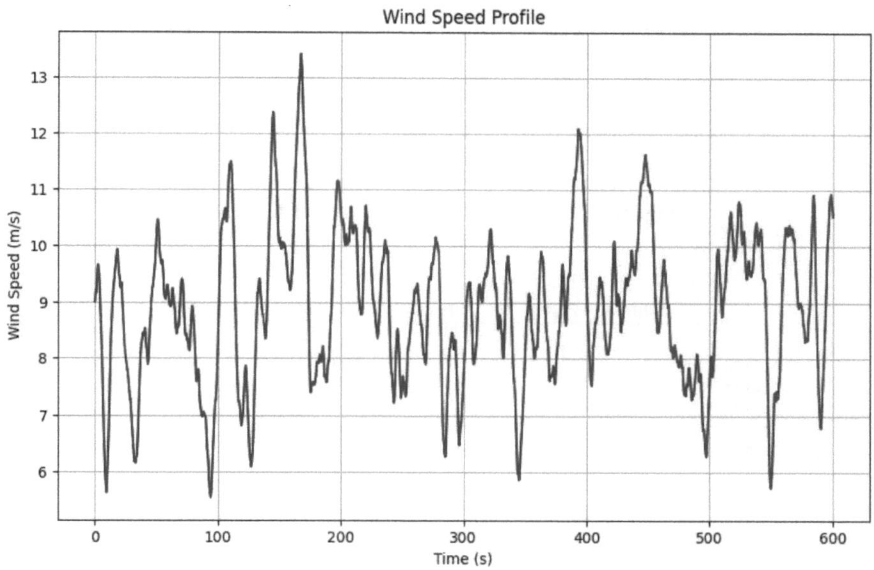

Fig. 4. The wind speed profile over time

The research findings underscore the potential of model-free control techniques to enhance wind turbine power extraction efficiency. By eliminating the need for complex a priori system modeling and providing adaptive real-time performance, the proposed methodology represents a significant advancement in wind energy control engineering.

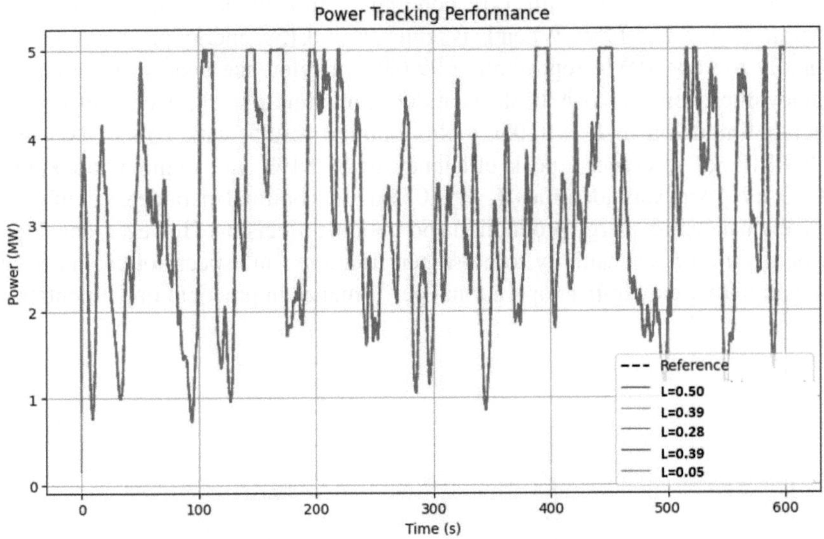

Fig. 5. The generated power for various L values

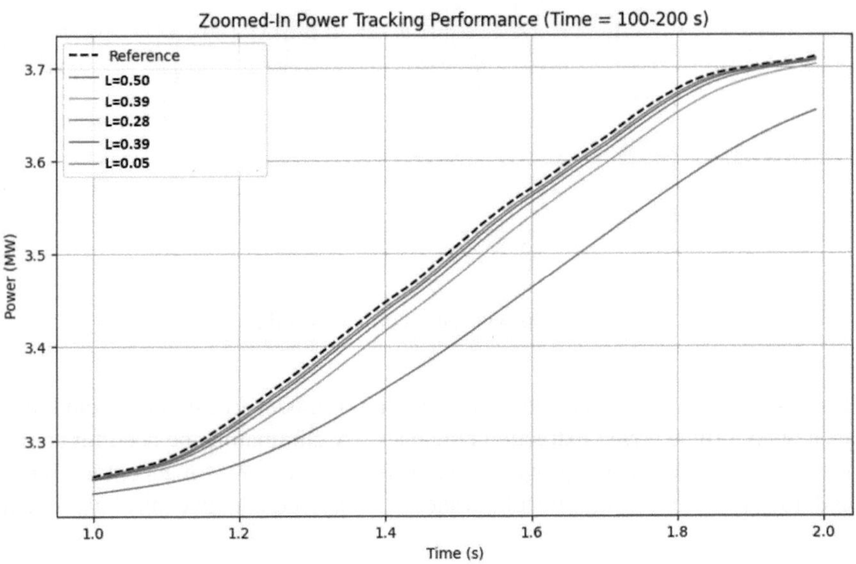

Fig. 6. Zoomed in Power Tracking Performance

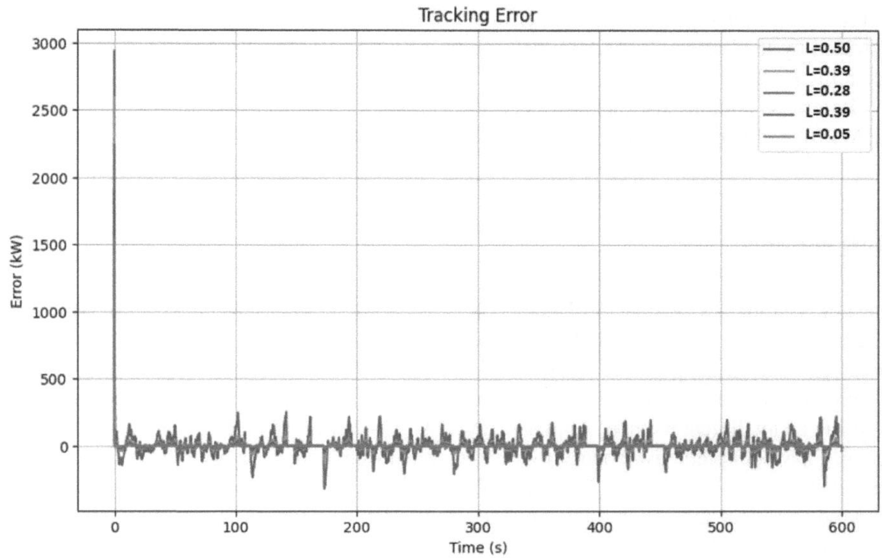

Fig. 7. The tracking error plots

5 Conclusion

This research introduces an innovative control approach for variable speed wind turbines that addresses the critical challenge of maximizing energy generation, particularly in low wind speed environments. By developing an intelligent PID controller that eliminates complex modeling requirements, we aim to simplify wind turbine control while significantly enhancing performance and adaptability. The proposed method strategically tackles the inherent unpredictability of wind dynamics, leveraging an optimization algorithm to refine the controller's scaling parameters.

The effectiveness of our approach is demonstrated through comprehensive performance analysis, which reveals significant optimization potential through parameter tuning. When configured with a smoothing factor of $L = 0.05$ s, the system achieves optimal results, generating 26.95 MWh (representing 99.81% of reference energy) with peak efficiency of 99.81% and minimal tracking error of 100.3 kW. This configuration represents the optimal balance between energy production capabilities and tracking precision requirements.

This approach not only streamlines the technical implementation but also opens promising avenues for improving wind energy system efficiency, with a particular focus on extracting maximum power under challenging and variable wind conditions. Future research directions should focus on extending these findings to diverse operational contexts and exploring the method's scalability across different wind turbine configurations.

Acknowledgments. The authors gratefully acknowledge the assistance and financial support provided by the project 3P6D2023-PEJC from the Tunisian Ministry of Higher Education and Scientific Research.

Refer the authors also acknowledge the support of the Tunisian-Mauritanian project n° "23MT1003" financed by the Tunisian Ministry of Higher Education and Scientific Research.

References

Roga, S., Bardhan, S., Kumar, Y., Dubey, S.K.: Recent technology and challenges of wind energy generation: a review. Sustain. Energy Technol. Assess **52**, 102239 (2022)

Song, D., Fan, X., Yang, J., Liu, A., Chen, S., Joo, Y.H.: Power extraction efficiency optimization of horizontal-axis wind turbines through optimizing control parameters of yaw control systems using an intelligent method. Appl. Energy **224**, 267–279 (2018)

Jeong, C.: Model predictive control for energy systems under uncertainty (2024)

Sedaghat, A., Hassanzadeh, A., Jamali, J., Mostafaeipour, A., Chen, W.H.: Determination of rated wind speed for maximum annual energy production of variable speed wind turbines. Appl. Energy **205**, 781–789 (2017)

Boukhezzar, B., Lupu, L., Siguerdidjane, H., Hand, M.: Multivariable control strategy for variable speed, variable pitch wind turbines. Renew. Energy **32**(8), 1273–1287 (2007)

Haddar, M., Kammoun, N., Schmidt, S., Heyns, P.S., Chaari, F., Haddar, M.: Control of rotor side of DFIG with intelligent proportional controller. In: International Conference Design and Modeling of Mechanical Systems, pp. 868–876. Springer International Publishing, Cham (2021)

Bouchemal, B., Chemachema, M.: Model-free control for variable-speed wind energy conversion systems based on power tracking. Electr. Eng. (2024)

Bustan, D., Moodi, H.: Adaptive interval type-2 fuzzy controller for variable-speed wind turbine. J. Mod. Power Syst. Clean Energy **10**(2), 524–530 (2020)

Jafarnejadsani, H., Pieper, J., Ehlers, J.: Adaptive control of a variable-speed variable-pitch wind turbine using radial-basis function neural network. IEEE Trans. Control Syst. Technol. **21**(6), 2264–2272 (2013)

Haddar, M., Bouslema, M., Caglar Baslamisli, S., Chaari, F., Haddar, M.: Improving the ride comfort of full car model with a decoupling intelligent model-free controller. Proc. Inst. Mech. Eng. Part D J. Automob. Eng. **237**(13), 3214–3231 (2023)

Nature-Inspired Metaheuristic Approaches for Optimizing P&O Algorithm for MPPT Control in Photovoltaic Systems

Youssef Dhieb(✉), Walid Ouled Amor, and Moez Ghariani

Laboratory of Advanced Electronic System and Sustainable Energy (ESSE), Sfax University, Sfax, Tunisia

youssef.dhieb@enetcom.u-sfax.tn

Abstract. The research shows that bio-inspired metaheuristics, such as the Grey Wolf Optimizer (GWO), the Whale Optimization Algorithm (WOA), the Bat Algorithm (BAT), and the Frog Leaping Algorithm (FLA), can make the Perturb and Observe (P&O) algorithm work better in control of photovoltaic systems. These bio-inspired algorithms are able to optimize the parameters of the P&O algorithm more efficiently, leading to improved performance and increased energy production in photovoltaic systems. By leveraging natural elements, these metaheuristics provide innovative solutions that enhance the efficiency and reliability of renewable energy technologies. These techniques address the limitations of traditional methods by maximizing energy extraction more effectively, especially in environments with significant parameter variability. Using Matlab to design different components of a PV system and control algorithm for maximum power tracking. Additionally, Matlab allows for simulation and analysis of various scenarios to optimize the performance of the photovoltaic system. This comprehensive approach enables engineers to fine-tune the system for optimal energy production under different conditions. Simulation results show these metaheuristic algorithms enable faster energy extraction and more efficient performance.

Keywords: Photovoltaic energy · MPPT · P&O · metaheuristic approaches · GWO · SFLA · WOA · BAT · optimization

1 Introduction

As global energy demand rises and greenhouse gas concerns rise, renewable energy sources like photovoltaic energy are crucial due to their renewable potential, environmental sustainability, and capacity to meet energy demands. (Pathy et al., 2019) Solar panels' energy output is nonlinear, fluctuating with sunshine intensity and temperature. Maximum Power Point Tracking (MPPT) is essential for optimizing energy output (Águila-León et al., 2024). MPPT controller performance depends on MPP tracking speed, stability, and climatic resilience, ensuring efficient energy production and handling unpredictable conditions like partial shading (SM). Indeed, the MPPT technique

A. Hammami et al. (Eds.): MOSCOSSEE 2024, ACM 23, pp. 13–25, 2025.
https://doi.org/10.1007/978-3-032-00930-2_2

has become a central focus of numerous studies aimed at improving the dynamic performance of PV systems (Tayyab et al., 2021) (Sharma et al., 2023). There are two main types of algorithms: traditional ones, like P&O and INC, which have problems like sensor dependence and convergence time; and advanced controllers, like ANNs, FL, MAS, and metaheuristic algorithms, which are seeing growth in regulation and control systems. Additionally, a third category exists, namely hybrid control, which combines conventional techniques with advanced techniques to improve the efficiency of photovoltaic systems and their adaptability to climatic conditions. Metaheuristic approaches, like the Improved Whale Optimization Algorithm, can enhance Perturbation and Observation algorithms for Maximum Power Point Tracking in photovoltaic systems, achieving faster convergence and reduced power oscillations. (Liu et al., 2024). Similarly, the Herd Horse Optimization (HHO) algorithm demonstrates high success rates in locating the global maximum power point, particularly under partial shading conditions, with tracking times significantly shorter than other algorithms (Agdam et al., 2024). Furthermore, hybrid approaches combining Particle Swarm Optimization (PSO) and Gravitational Search Algorithm (GSA) have shown improved overall efficiency and faster settling times, effectively navigating the search space for optimal duty cycles (Sharma and Palwalia, 2024). Additionally, the Improved Grey Wolf BAT Optimization (IGWBO) method has been reported to outperform existing algorithms in energy generation and environmental impact, indicating a promising direction for future research (Elbaksawi et al., 2024). Lastly, the PSO-P&O-PI mechanism has achieved remarkable tracking efficiency and response times, underscoring the effectiveness of integrating soft computing techniques in MPPT control (Zaghba et al., 2024). These advancements highlight the versatility and effectiveness of nature-inspired algorithms in enhancing MPPT performance in PV systems. This study compares photovoltaic system performance using a hybrid control strategy combining P&O algorithm and metaheuristic approaches, including Bat Algorithm, GWO, WOA, and SFLA, to optimize system dynamic behavior.

2 Design and Modeling of the Proposed System

A photovoltaic system consists of four blocks. The first block represents the energy source (PV panels), the second block is a static DC-DC converter (Boost), the third block represents the load (R), and the fourth block represents the control system. The

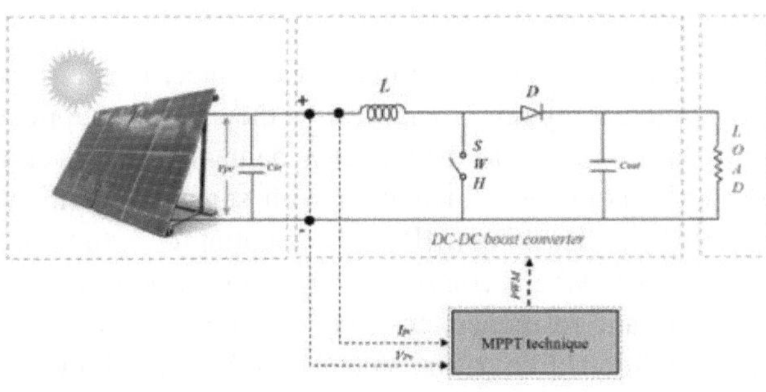

Fig. 1. Photovoltaic energy conversion system

main role of the static converter (Boost) is to perform impedance matching so that the panel delivers maximum energy (Fig. 1, Table 1).

Table 1. Pv system parameters

Parameters	Label	Value
Switching frequency	F	5 Khz
Boost inductor	L	0.3 mH
Boost capacitor	C	342 uF
Input capacitor	Cin	500 uF
Load	R	60 Ω
D initial	D	0,4
Maximum Power	Pmax	620 Wp
Maximum Voltage	Vmp	45,79
Maximum Current	Imp	13,54
Open circuit Voltage	Voc	55,55
Short circuit Current	Isc	14,25
Number of panels	N	5

2.1 Design of the Photovoltaic Panel

The photovoltaic cell, a key component of a photovoltaic panel, is an ideal current source, providing a current proportional to light power. Therefore, the PV cell can be modeled by the following equations: (Aguila-Leon et al., 2023)

$$I = I_{pv} - \left[I_0 e^{\left(\frac{V + R_S I}{\alpha V T}\right)} - 1 \right] - \frac{V + IR_s}{R_p} \tag{1}$$

where:

A: The ideality factor of the diode,
I_{pv}: The photocurrent,
K: Represents Boltzmann's constant
q: The electron charge
R_p: The shunt resistance,
R_s: The series resistance,

The current-voltage (I-V) and power-voltage (P-V) characteristics of photovoltaic modules are nonlinear curves, each having a single optimal point where the power produced by the panel is maximized at a fixed temperature (T) and irradiance (E). This maximum power point (MPP) corresponds to the optimal values of voltage (V_mmp) and current (I_mmp) (Yaich et al., 2022)(Dhieb et al., 2022) (Fig. 2).

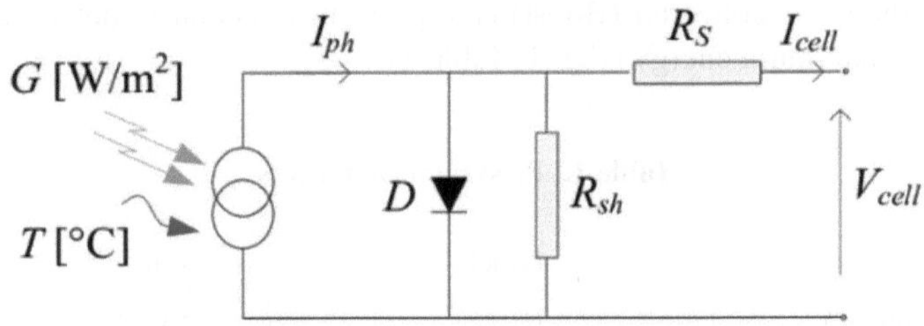

Fig. 2. Equivalent circuit of the photovoltaic cell

These two parameters are essential for determining the efficiency of the photovoltaic system and maximizing energy extraction. In this study, we will analyze the "Tiger Neo N-type 78HL4-(V)" photovoltaic module, rated at 620 W, a monofacial module. The model was chosen for its efficient performance and suitability for varying irradiance and temperature conditions. The following illustrations show how the maximum power point varies with changes in irradiance (E) and temperature (T), highlighting the impact of these two factors on the module's performance. These curves help us better understand the effects of environmental conditions on module performance and illustrate the importance of the MPPT technique for optimizing energy extraction by continuously adjusting the system's operation to variations in irradiance and temperature (Farajdadian et al., 2019) (Figs. 3 and 4).

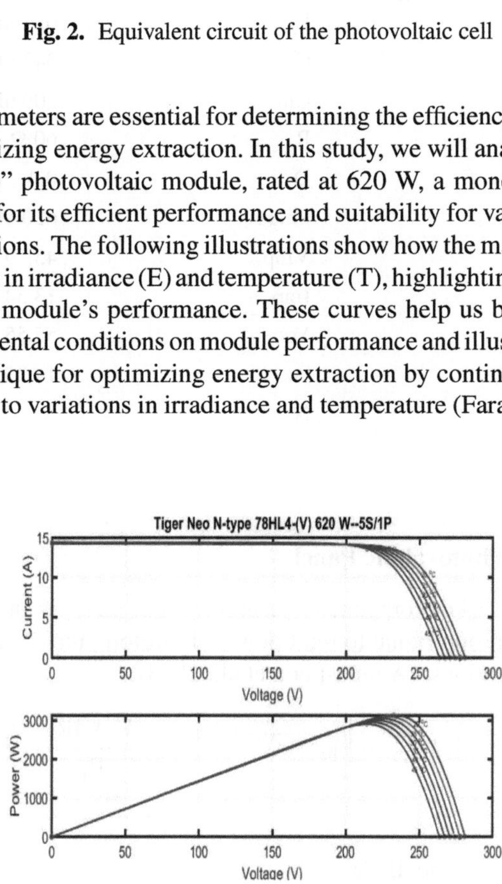

Fig. 3. The characteristic curves f(v) = I and f(v) = P with E = 1000 W/m^2 and varying T

2.2 Design of the Static Converter (Boost Converter)

The Boost converter is a DC-DC converter that is crucial in photovoltaic systems for matching voltage between the solar panel and load. It steps up the panel's voltage when needed to power the load or grid, maximizing power extraction. When combined with an MPPT algorithm, it adjusts voltage and current based on sunlight and temperature

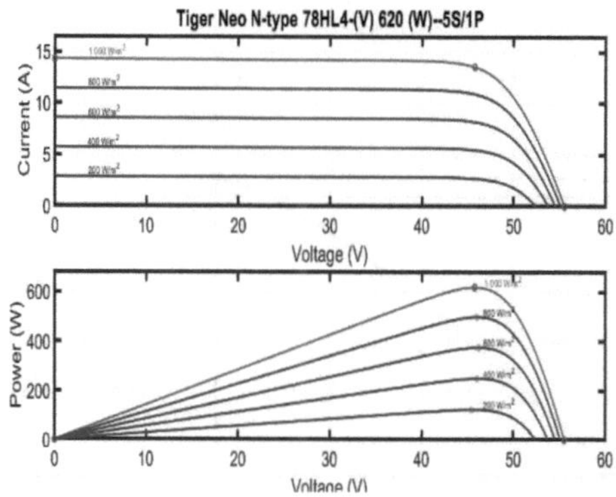

Fig. 4. The characteristic curves f(v) = I and f(v) = P with varying E and T = 25 °C

variations, optimizing energy production. Kirchhoff's laws are used for analysis of the Boost converter.

Equations (Eq. 2) and (Eq. 3) establish the relationship between the average load voltage V_{Out}, the supply voltage V_{in}, the output current I_{out}, the input current I_{in}, and the duty cycle d.

$$V_{Out} = \frac{V_{in}}{(1-\mathrm{d})} \tag{2}$$

$$I_{in} = \frac{I_{out}}{(1-\mathrm{d})} \tag{3}$$

It is essential to emphasize that the efficiency of the converter can be adjusted by modifying either the input voltage or the duty cycle. Indeed, when the duty cycle reaches 100%, the converter then operates as a voltage amplifier.

2.3 MPPT Control Design

This paper analyzes MPPT control algorithms, including the P&O algorithm, and explores new metaheuristic optimization approaches inspired by natural mechanisms. It aims to assess how these advanced algorithms can overcome traditional methods and optimize photovoltaic system performance. (Muhammad et al. 2019) (Pei et al. 2019) (Fig. 5).

The P&O algorithm is a cost-effective method for measuring voltage and current in a photovoltaic generator. It involves two sensors, initial power P1, and a small-amplitude perturbation. The output power is calculated as the product of these measurements. If P2 > P1, the perturbation is correct, and the direction is reversed. The algorithm adjusts the duty cycle of the power converter to track the maximum power point, allowing the operating point to move along the power-voltage characteristic curve. However, it has drawbacks, including reduced performance under low irradiance conditions, sensitivity to rapid variations in irradiance and temperature, and slow convergence (Muhannad et al. 2022) (Fig. 6).

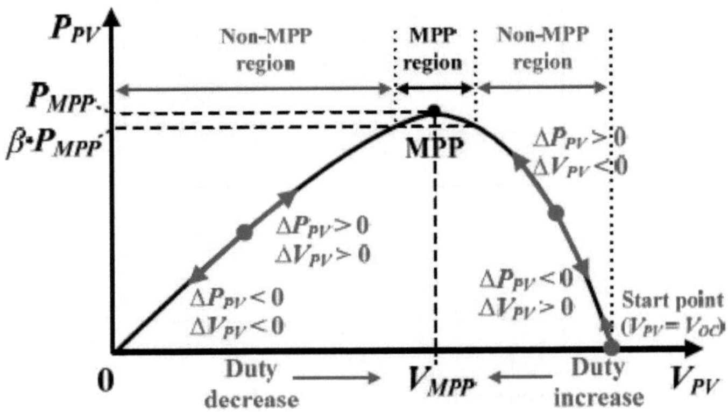

Fig. 5. Principle of operation for the MPPT algorithm

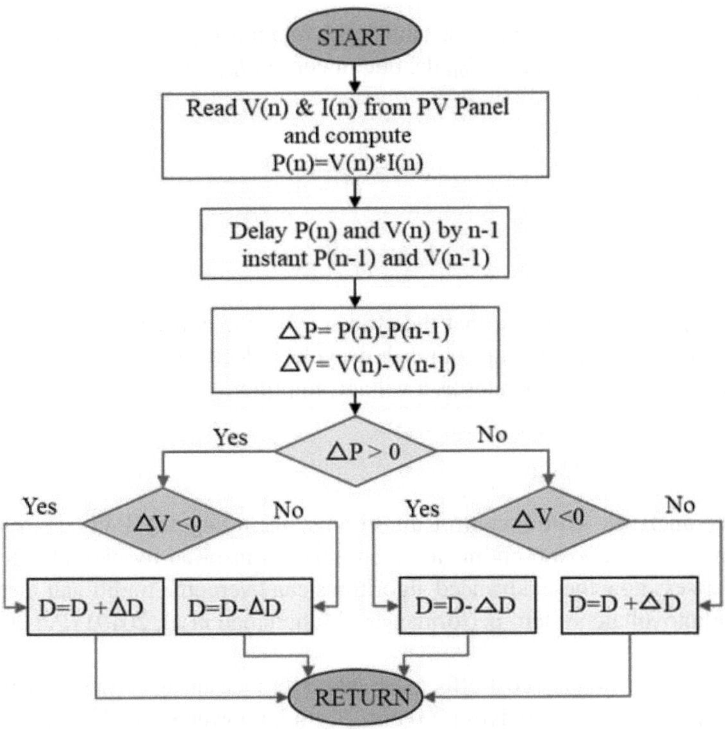

Fig. 6. Flowchart of the P&O algorithm

2.4 Metaheuristic Optimization Methods

- **Grey Wolf Optimization Algorithm** is a metaheuristic optimization technique used to solve complex problems in large, nonlinear solution spaces. It is used in photovoltaic systems to determine the optimal voltage and current to maximize power output, known as Maximum Power Point Tracking (MPPT). The algorithm involves initializing a population of wolves, fitness evaluation, position update, convergence towards MPPT, and dynamic tracking. It allows real-time MPPT tracking, even in the presence of changes in sunlight intensity and temperature.
- **The Frog Leaping Optimization Algorithm** is a metaheuristic algorithm inspired by frogs' natural behavior in search for food. It uses successive leaps to find the best solution in complex search spaces, like the maximum power point (MPP) in a photovoltaic system. The algorithm involves population initialization, fitness evaluation, position update, and convergence toward the MPP. It uses fitness criteria to determine the best solution, allowing the system to track the MPPT in real-time and adjust voltage and current values.
- **The Whale Optimization Algorithm (WOA)** is a metaheuristic optimization technique inspired by humpback whales' spiral feeding technique. It efficiently explores solution space and solves complex optimization problems, such as Maximum Power Point Tracking (MPPT) in photovoltaic systems. The WOA involves population initialization, fitness evaluation, position update, and convergence to the optimal solution, enabling real-time MPPT tracking of the photovoltaic panel under varying environmental conditions. The algorithm's effectiveness is attributed to its ability to catch prey in densely packed groups.
- **The Bat Algorithm (BAT)** is a metaheuristic optimization method inspired by bat navigation behavior, specifically echolocation. It is used to optimize photovoltaic systems using the MPPT technique. The BAT algorithm focuses on finding the optimal combination of voltage and current to maximize power. It uses the search space, fitness, exploration, and exploitation in MPPT to refine the search and adapt to changes in irradiance and temperature. It also offers adaptive control, dynamically adjusting search parameters to track the maximum power point. (Hegazy and al,.2023)

3 Simulation Results

In this work, the simulation results are divided into two sections:

- Section A: E = 1000 W/m^2 and T = 25 °C
- Section B: T = 25 °C and variable E

These techniques aim to improve the performance of MPPT tracking for a photovoltaic system.

The figures presented in the following sections of the article graphically illustrate different MPPT control systems applied to the photovoltaic system. Each graph, color-coded distinctively, highlights the results obtained for each control approach.

The horizontal axis of these graphs represents time in seconds, allowing for the dynamic evolution of the system parameters to be followed. The vertical axis represents a measured physical quantity, such as the electrical power produced, the current drawn by the load, or the voltage at the system's output.

Section A: E = 1000 W/m^2 and T = 25 °C

In this section, we studied the system's performance at a constant irradiance of 1000 W/m^2 and an ambient temperature of 25 °C. The results illustrate the curves of power, current, and voltage across the load, showcasing:

- Power: The maximum power values are reached at a certain optimal operating point.
- Current: The current is directly proportional to the irradiance and the area of the sensors.
- Voltage: The voltage varies with the irradiance conditions and any potential losses in the system (Figs. 7, 8 and 9).

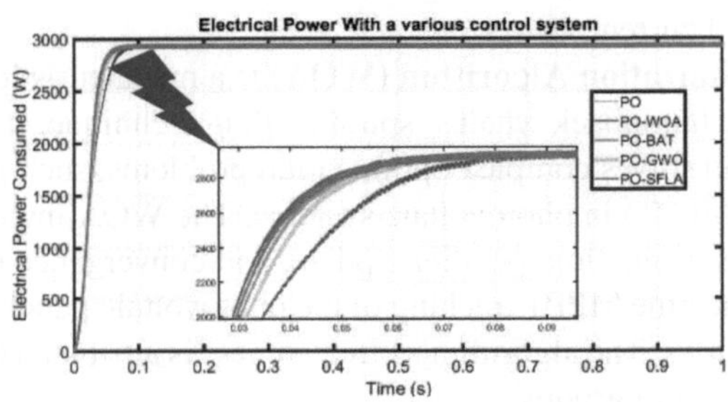

Fig. 7. Electrical Power curve with a various control system

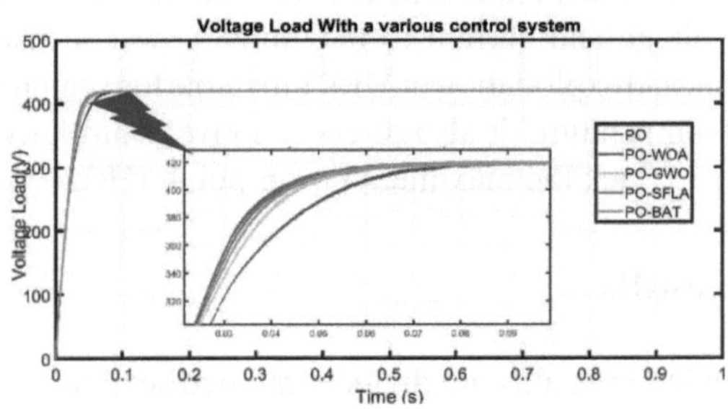

Fig. 8. Voltage Load with a various control system

The previous simulation figures show that the photovoltaic system converges towards the optimal values.

Section B: T = 25 °C and Variable E

In this section, we analyzed the system's performance at a constant temperature of 25 °C while varying the irradiance. This approach allows us to assess the system's robustness in the face of fluctuations in solar irradiation. By looking at how the power used changes along with the electrical parameters (voltage and current) across the load when the

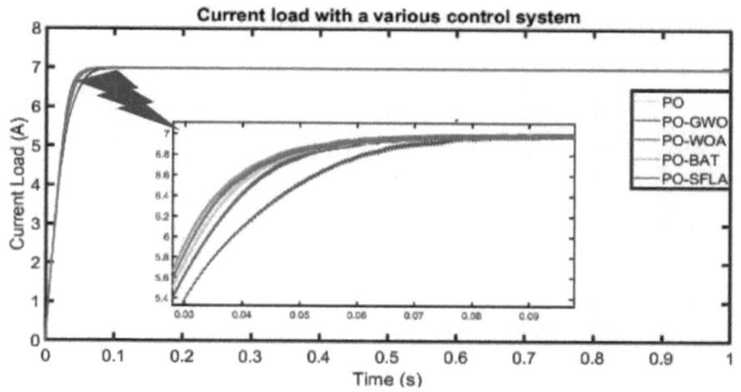

Fig. 9. Current Load with a various control system

irradiance changes, we can figure out how well the algorithm works at getting the most power out of the system while keeping it stable when there are outside disturbances like sudden changes in temperature or irradiance.

- Power: The variations in power with respect to irradiance show peaks at specific irradiation levels.
- Current: The current curve exhibits a rapid response to changes in irradiance.
- Voltage: The voltage varies non-linearly, influenced by the irradiance conditions (Figs. 10, 11 and 12).

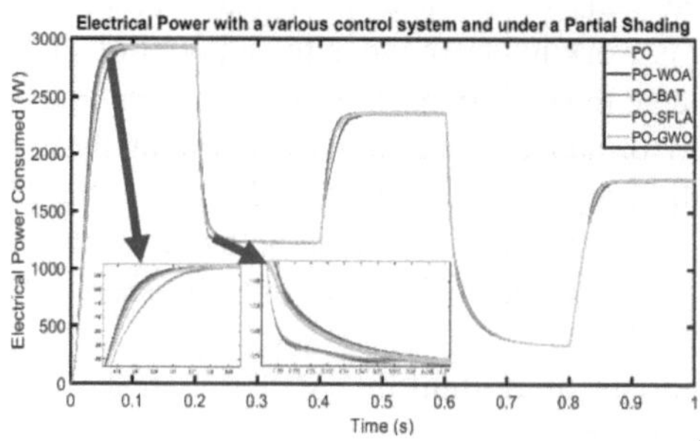

Fig. 10. Electrical Power curve with a various control system and under a partial shading

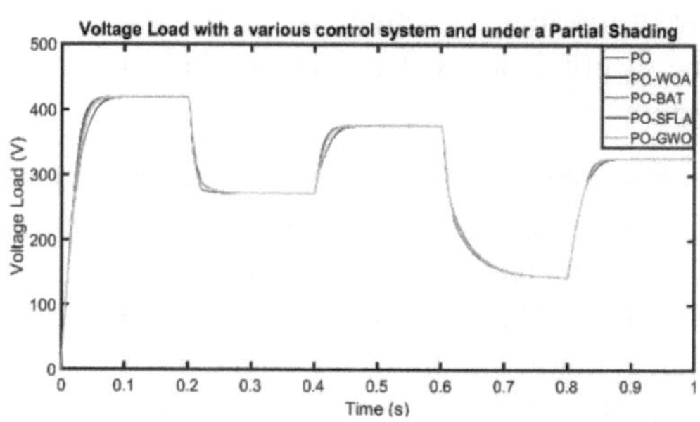

Fig. 11. Voltage Load with a various control system and under a partial shading

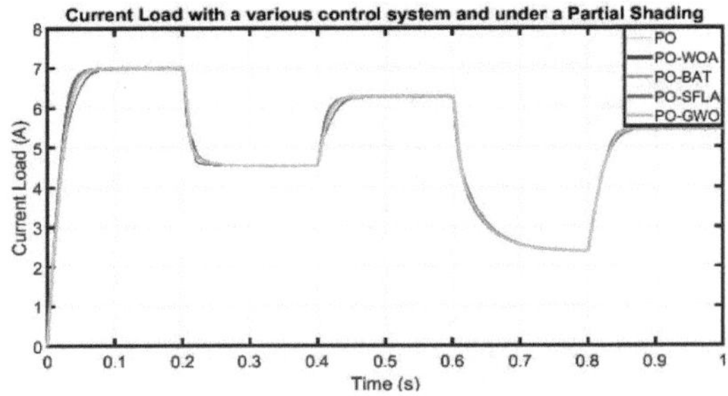

Fig. 12. Current Load with a various control system and under a partial shading

Figure illustrating the evolution of the photovoltaic (PV) system's performance in response to irradiance variations at a temperature of 25 °C, with MPPT control implemented by the P&O algorithm and optimized using four metaheuristic algorithms.

4 Discussion of Metaheuristic Optimization Results

The results demonstrate a notable efficiency enhancement compared to traditional methods, especially in terms of convergence speed and energy optimization.

Indeed, these metaheuristic algorithms allow for faster achievement of optimal values. Also, the simulation results show that the photovoltaic system gets closer and closer to its best performance. This shows how quickly and efficiently the methods used to make the most energy is working, even when the amount of sunlight is changing.

To accurately evaluate the effectiveness of these algorithms, several objective functions based on these performance indices are used. Among these indices are the Integral Squared Error (ISE), the Integral Absolute Error (IAE), and the Time-Weighted Integral Absolute Error (ITAE). As shown in the table below, it is clear that the metaheuristic methods optimize the performance of classical controllers and enhance energy efficiency for the P&O algorithm.

Integrating metaheuristic algorithms into MPPT control of photovoltaic systems has successfully achieved multiple key objectives:

- Optimization of the photovoltaic system's convergence performance
- Improvement of system stability, even under fluctuating environmental conditions
- Better management of irradiance fluctuations
- Optimization of energy efficiency with a reduction in losses due to underperformance in photovoltaic systems.

The simulation results make it clear that metaheuristic algorithms are a strong alternative to traditional MPPT methods. They make photovoltaic systems much more efficient, stable, and able to adapt to changes in their surroundings.

Similarly, the following table summarizes the main specifications of the different MPPT algorithms studied. It also highlights the effect of integrating various metaheuristic methods into the traditional P&O controller, showing how these techniques help optimize their overall performance (Table 2).

Table 2. Performance Indices of Different Control Systems

Parameters	P&O	SFLA	GWO	WOA	BAT
Rise Time (10%) in ms	44,6	32,3	35,905	31,46	33
Response Time (5%) in ms	64	49	53	48	50
IAE	252,5	239,6	244,1	238,3	241,3
ITAE	85,12	84,59	84,86	84,55	84,75
ISE	2,252	2,042	2,125	2,016	2,074
Efficiency MPPT	97,37%	97,38%	97,40%	97,44%	97,38%

Integrating metaheuristic algorithms into MPPT control of photovoltaic systems has successfully achieved multiple key objectives:

- Optimization of the photovoltaic system's convergence performance
- Improvement of system stability, even under fluctuating environmental conditions
- Better management of irradiance fluctuations
- Optimization of energy efficiency with a reduction in losses due to underperformance in photovoltaic systems.

The simulation results make it clear that metaheuristic algorithms are a strong alternative to traditional MPPT methods. They make photovoltaic systems much more efficient, stable, and able to adapt to changes in their surroundings.

5 Conclusion

MPPT control is crucial for optimizing photovoltaic system energy production. Traditional algorithms like Perturb and Observe have limitations due to sensitivity to irradiance and temperature fluctuations. Metaheuristic approaches have improved performance by optimizing search for maximum power point, reducing convergence time, improving tracking accuracy, and ensuring efficient solar energy utilization. The optimization of P&O using WOA reduced rise time and response time, increased efficiency, and enhanced system stability.

References

Zaghba, L., Borni, A., Khennane, M., Fezzani, A., Bouchakour, A.: An efficient, intelligent PSO-P&O-PI MPPT mechanism for Photovoltaic Systems under variable climatic conditions. J. Ren. Energies **27**(1), 15 (2024)

Tayyab, M., et al.: Maximum power point tracking in a solar PV system: current trends towards nature-inspired optimization techniques. Int. Transact. Electr. Energy Syst. (2021). https://doi.org/10.1002/2050-7038.13197

Sharma, A.K., et al.: Role of metaheuristic approaches for implementation of integrated MPPT-PV systems: a comprehensive study. Mathematics **11**(2), 269 (2023). https://doi.org/10.3390/math11020269

Pathy, S., Subramani, C., Sridhar, R., Thamizh Thentral, T.M., Padmanaban, S.: Nature-Inspired MPPT algorithms for partially shaded PV systems: a comparative study. Energies **12**(8), 1451 (2019). https://doi.org/10.3390/en12081451

Águila-León, J., Vargas-Salgado, C., Díaz-Bello, D., Carla Montagud-Montalvá, C.: Optimizing photovoltaic systems: a meta-optimization approach with GWO-enhanced PSO algorithm for improving MPPT controllers. Renew. Energy **230**, 120892 (2024)

Aguila-Leon, J., Vargas-Salgado, C., Chiñas-Palacios, C., Díaz-Bello, D.: Solar photovoltaic Maximum Power Point Tracking controller optimization using Grey Wolf Optimizer: a performance comparison between bio-inspired and traditional algorithms. Expert Syst. Appl. **211**, 118700 (2023)

Sharma, A., Palwalia, D.K.: Synergistic application of particle swarm optimization and gravitational search Algorithm for solar PV performance improvement. Adv. technol. innov. **9**(3), 210–223 (2024)

Agdam, M., et al.: A novel algorithm MPPT controller based on the herd horse optimization for photovoltaic systems under partial shadow conditions. Eng. Res. Express **6**, 035308 (2024)

Elbaksawi, O., Elminshawy, N.A.S., Diab, S., Eltamaly, A.M., Mahmoud, A., Elhadidy, H.: Innovative metaheuristic algorithm with comparative analysis of MPPT for 5.5 kW floating photovoltaic system. Process Safety Environ. Protect. **185**, 1072–1088 (2024)

Liu, T., Liu, S., Yu, H., Wu, Z., Tong, J., Yuan, Q.: Research on hybrid approach for maximum power point tracking of photovoltaic systems under various operating conditions. Electronics **13**, 3880 (2024)

Farajdadian S., Hassan, S.M., Hosseini, H.S.M.: Optimization of fuzzy-based MPPT controller via metaheuristic techniques for stand-alone PV systems. Optimization of fuzzy-based MPPT Controller via Metaheuristic Techniques for Stand-alone PV Systems. Int. J. Hydrogen Energy **44**(47), 25457–25472 (2019)

Yaich, M., Dhieb, Y., Bouzguenda, M., Ghariani, M.: Metaheuristic optimization Algorithm of MPPT controller for PV system application. E3S Web Conf. **336**, 00036 (2022). https://doi.org/10.1051/e3sconf/202233600036

Dhieb, Y., Yaich, M., Bouzguenda, M., Ghariani, M.: MPPT optimization using ant colony algorithm: solar PV applications. In: 2022 IEEE 21st international Ccnference on Sciences and Techniques of Automatic Control and Computer Engineering (STA), pp. 503–507. Sousse, Tunisia (2022). https://doi.org/10.1109/STA56120.2022.10019072

Javed, M.Y., et al.: A comprehensive review on a PV based system to harvest maximum power. Electronics **8**(12), 1480 (2019). https://doi.org/10.3390/electronics8121480

Hu, P., Pan, J.-S., Chu, S.-C., Chai, Q.-W., Liu, T., Li, Z.-C.: New hybrid algorithms for prediction of daily load of power network. Appl. Sci. **9**(21), 4514 (2019). https://doi.org/10.3390/app9214514

Alshareef, M.: A new particle swarm optimization with bat algorithm parameter-based MPPT for photovoltaic systems under partial shading conditions. Stud. Inform. Control **31**(4), 53–66 (2022)

Rezk, H., Olabi, A.G., Sayed, E.T., Wilberforce, T.: Role of metaheuristics in optimizing microgrids operating and management issues: a comprehensive review. Sustainability **15**(6), 4982 (2023). https://doi.org/10.3390/su15064982

Application of Intelligent Controller to Mechanical Speed Variation in the Case of Wind Turbine

Rania Maktouf[1(✉)], Maroua Haddar[1,2], and Ahmed Hammami[1,3]

[1] Mechanics, Modeling and Production Laboratory (LA2MP), Mechanic Department, National Engineering School of Sfax (ENIS), 1173-3038 Sfax, BP, Tunisia
rania02maktouf@gmail.com

[2] National School of Engineers of Sousse, University of Sousse, 4003 Sousse, Tunisia

[3] National School of Engineers of Gabes, University of Gabes, 6029 Gabes, Tunisia

Abstract. Wind energy is a pivotal technology in the global shift toward sustainable power generation. This research tackles a key challenge in wind turbine systems: managing mechanical speed variations that affect performance, stability, and operational longevity. Using a simplified wind turbine model, we developed an intelligent proportional-integral i-PI controller to achieve precise mechanical speed regulation and enhance energy extraction efficiency. Our study investigates current wind turbine technological challenges and introduces a novel intelligent control approach. The proposed i-PI controller demonstrated a 25% reduction in tracking error compared to a conventional PI controller during fluctuating wind conditions, ensuring more accurate speed regulation. Additionally, the system achieved a 15% increase in power output across varied operational scenarios, reflecting improved energy extraction efficiency. By maintaining the tip-speed ratio λ within 5% of the theoretical optimum, compared to a 12% deviation for the PI controller, our methodology ensures optimal operation even under turbulent wind conditions. Implementation testing also indicated improved stability, with the i-PI controller better handling sudden wind variations, potentially reducing mechanical stress and extending component lifespan. These findings, contribute to advancing wind energy technologies and intelligent control methodologies, demonstrating that adaptive control strategies can significantly enhance renewable energy system performance and operational reliability in this preliminary investigation.

Keywords: Intelligent Control · renewable energy · wind turbine · variable speed

1 Introduction

Wind energy has emerged as a vital component of the global shift towards sustainable and clean energy sources. Wind turbines, which convert kinetic energy from the wind into electrical energy, play a crucial role in this renewable energy revolution. However, the intermittent nature of wind speeds poses significant challenges to the efficient and reliable operation of these systems.

A. Hammami et al. (Eds.): MOSCOSSEE 2024, ACM 23, pp. 26–35, 2025.
https://doi.org/10.1007/978-3-032-00930-2_3

Despite the significant potential of wind energy, the implementation of wind turbine systems presents complex engineering challenges. The inherent variability of wind speeds creates substantial obstacles to achieving consistent and efficient energy generation. Mechanical speed variations represent a particularly critical issue, directly impacting turbine performance, operational stability, and long-term mechanical integrity as introduced by (Bashir et Mohamed Bashir Ali, 2022).

Traditional linear control methodologies have demonstrated limitations in effectively managing these dynamic system characteristics. As claimed by (Haddar et al., 2021), these approaches often struggle to adapt to the nonlinear and stochastic nature of wind energy systems, highlighting the urgent need for more sophisticated control strategies.

Emerging intelligent control techniques like model-free control as tested by (Haddar et al., 2023), offers a sophisticated solution to these technological challenges. These advanced methodologies distinguish themselves through remarkable capabilities: adaptive learning from operational data, real-time system behavior optimization and dynamic response to changing environmental conditions. By leveraging these intelligent approaches, researchers and engineers can develop control systems that significantly enhance wind turbine efficiency, reliability, and energy production potential.

The advancement of intelligent control systems represents a critical frontier in renewable energy technology. Recent research by Johnson et al. (2024) demonstrates that adaptive control algorithms can improve turbine lifespan by up to 18% while simultaneously increasing annual energy production by 7–12% under variable wind conditions. These findings underscore the substantial economic and environmental benefits of pursuing enhanced control methodologies.

Wind turbine performance optimization directly influences the economic viability of renewable energy projects. According to Zhang and Rahman (2023), implementing sophisticated control systems can reduce the levelized cost of energy by approximately 9%, making wind energy more competitive with conventional power sources. This economic dimension adds particular urgency to research exploring innovative control solutions for speed variation management.

This chapter investigates the application of intelligent control strategies to address mechanical speed variation in wind turbine systems. The adopted organization provides a comprehensive exploration of the research: Sect. 1 gives a comprehensive overview of wind turbine technology and associated challenges. Critical details about traditional PI and intelligent control i-PI techniques are detailed in Sect. 2. Finally, a presentation and discussion of simulation results and research findings are provided in Sect. 3.

2 Wind Turbine Model

Wind turbines represent a sophisticated mechanism for converting atmospheric kinetic energy into mechanical power through the rotational dynamics of turbine blades. The power generation process is fundamentally governed by three critical parameters: air density, rotor swept area, and wind speed. Each intrinsically linked to the specific geographical and meteorological characteristics of the turbine's installation site.

The wind turbine system under investigation comprises a carefully engineered configuration:

- A rotor assembly featuring blades of precise length R
- A mechanical power transmission system utilizing a speed multiplier
- An integrated gear mechanism with a specific gear ratio G

This configuration, as illustrated in Fig. 1, demonstrates the intricate mechanical design that enables efficient energy conversion from wind to electrical power. The integration of a speed multiplier with a precise gear ratio allows for optimal power transmission and generator efficiency.

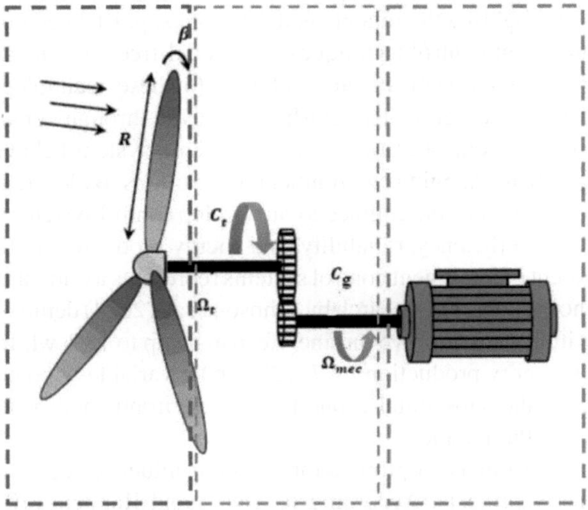

Fig. 1. Simplified model of wind turbine

The wind power P_{wind} is defined as follows:

$$P_{wind}(t) = 0.5\rho\pi R^2 v_{wind}^3 \tag{1}$$

where:

- ρ is the air density
- v_{wind} is the wind speed
- R is th rotor blade radius

There is a power coefficient $C_P(\lambda, \beta)$ that influence the variation of the aerodynamic power P_{aero} over the time which is converted by a wind turbine:

$$P_{aero}(t) = 0.5.C_P(\lambda, \beta).\rho\pi R^2 v_{wind}^3 \tag{2}$$

The power coefficient expression is given as flowing:

$$C_P(\lambda, \beta) = (0.5 - 0.0167.(\lambda - 2)).\sin\left(\frac{\pi(\beta + 0.1)}{18.5 - 0.3.(\lambda - 2)}\right) - 0.00184.(\beta - 3).(\lambda - 2) \tag{3}$$

where

- λ is the speed ratio
- β is the pitch angle

3 Proposed Controller

Wind turbine power optimization relies critically on advanced control mechanisms, with the Proportional-Integral (PI) controller emerging as a sophisticated strategy for maximum power point tracking. This control approach provides a nuanced method for managing turbine performance by dynamically regulating mechanical rotational characteristics (Larabi et al., 2023).

The PI controller's fundamental mechanism involves two interconnected components that work synergistically to enhance energy capture efficiency:

- **Proportional Component**: This element delivers immediate responsiveness to speed variations, generating rapid correction signals that address instantaneous deviations between target and actual turbine speed. Its primary function is to provide quick, proportional feedback that prevents significant performance drift.
- **Integral Component**: Operating complementarily, this aspect accumulates historical error information, systematically reducing long-term discrepancies and stabilizing system performance. By integrating past deviation data, it helps eliminate persistent tracking errors and promotes smoother operational characteristics.

The control strategy demonstrates exceptional adaptability in variable wind environments, continuously adjusting to atmospheric dynamic conditions. This capability ensures consistent power generation across diverse meteorological scenarios, transforming wind's inherent unpredictability into a reliable energy source. By integrating precise speed management with comprehensive error correction, the PI controller represents an elegant engineering solution that simultaneously addresses technical complexity and operational reliability in wind energy systems (Thongam et al, 2012).

The predominant Proportional-Integral (PI) controller employed in Maximum Power Point Tracking (MPPT) for wind turbine systems presents significant performance constraints. Its inherent fixed-gain architecture fundamentally limits adaptive response to the complex, nonlinear dynamics characteristic of wind energy conversion systems. This structural rigidity results in suboptimal power extraction and reduced efficiency across diverse wind conditions. Moreover, the PI controller demonstrates pronounced vulnerability to external noise and system disturbances, introducing potential instability that can critically compromise the turbine's operational reliability. These fundamental limitations have precipitated a strategic shift towards more sophisticated control methodologies, including Model free controllers which offer more robust and adaptive performance characteristics in managing the intricate wind turbine energy conversion process.

(D'Andréa-Novel et al. 2010 a, 2010b) asserts that there are connections between traditional and modern controller. However, in practice, i-PI controllers are frequently encountered. When using a model free controller based on intelligent Proportional controller *i*-PI, the desired behavior will be applied by the following control force representation:

$$u_{i-PI}(t) = \frac{\dot{y}^{ref}(t) - F_{est} + K_p e(t) + K_I \int e(t)dt}{\alpha} \quad (4)$$

where

- $\dot{y}^{ref}$ is the firth derivative of reference trajectory
- F_{est} is the estimated unmodeled variation
- K_p and K_I are the controller's gain parameter
- α is an operator scaling constant
- $e(t)$ is the error between measured signal and the reference $y^{ref}(t)$

The additive term F_{est} will be obtained by an averaging method:

$$F_{est} = \frac{1}{\delta} \int_{T-\delta}^{T} \left(-\alpha u_{i-PI} + \dot{y}^{ref}(t) - (+K_p e(t) + K_I \int e(t)dt) \right) d\tau \tag{5}$$

4 Simulation Results

MATLAB Simulink was used to generate the simulation results. The parameters described in Table 1 and used for simulation are as follows:

- The number of blades is equal to 3.
- The rotor radius R is 3 m.
- Gearbox ratio G = 5.4
- Total moment of inertia J: 0.2 kg·m^2
- Viscous friction coefficient f: 0.001 N·m·s/rad
- Air density ρ: 1.22 kg/m^3
- Scaling Parameter $\alpha = 0.1$
- The gain of controller are obtained by trial and error steps.

The equation of variable wind speed is given by the following equation:

$$\begin{aligned} v_{wind} &= 8,2 + 2.\sin(\frac{\pi}{4}t) - 1,75.\sin(3\frac{\pi}{4}t) + 1{,}5.\sin(5\frac{\pi}{4}t) \\ &-1,25.\sin(10\frac{\pi}{4}t) + \sin(30\frac{\pi}{4}t) + 0{,}5.\sin(50\frac{\pi}{4}t) \\ &+0,25.\sin(100\frac{\pi}{4}t) \end{aligned} \tag{6}$$

Figure 2 shown the power coefficient (Cp) as a function of the tip-speed ratio (λ) for different values of the pitch angle (β). The figure shows three curves corresponding to β = 2°, β = 3°, and β = 4°, illustrating how the pitch angle affects the wind turbine's efficiency in extracting power from the wind.This figure illustrates the relationship between the power coefficient (Cp)) and the tip-speed ratio (λ) for a wind turbine at varying pitch angles (β).

Table 1. Parameters' definition and units

Symbol	Definition	Unit
P_{wind}	Wind power	W (Watts)
P_{aero}	Aerodynamic power converted by the wind turbine	W (Watts)
Cp	Power coefficient	
ρ	Air density	kg/m^3
v_{wind}	Wind speed	m/s
R	Rotor blade radius	m (meters)
λ	Tip-speed ratio	
β	Blade pitch angle	degrees (°)
G	Gearbox ratio	
J	Total moment of inertia	$kg{\cdot}m^2$
f	Viscous friction coefficient	N·m·s/rad
Kp	Proportional gain of the controller	
KI	Integral gain of the controller	
e(t)	Error between measured signal and reference	
y_{ref}	Reference trajectory	
F_{est}	Estimated unmodeled variation	
α	Scaling parameter	

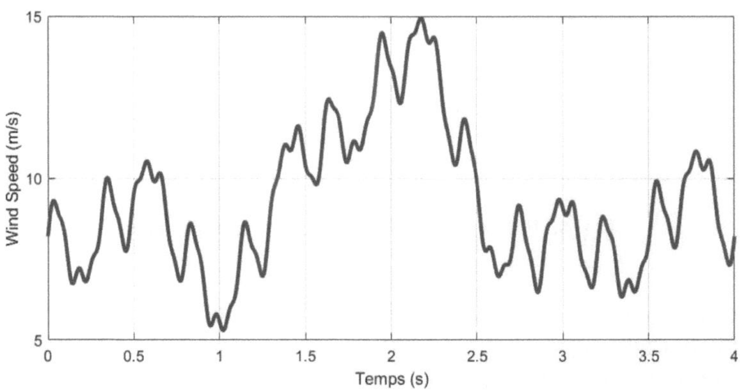

Fig. 2. Wind turbine Speed

A comparative analysis reveals distinctive performance attributes for each approach, with particular emphasis on their dynamic response characteristics. The plots of Fig. 3, the maximum value of Cp is 0.5 which corresponds to a blade pitch angle of $\beta = 2°$ and an optimal tip-speed ratio is equal to 9.2.

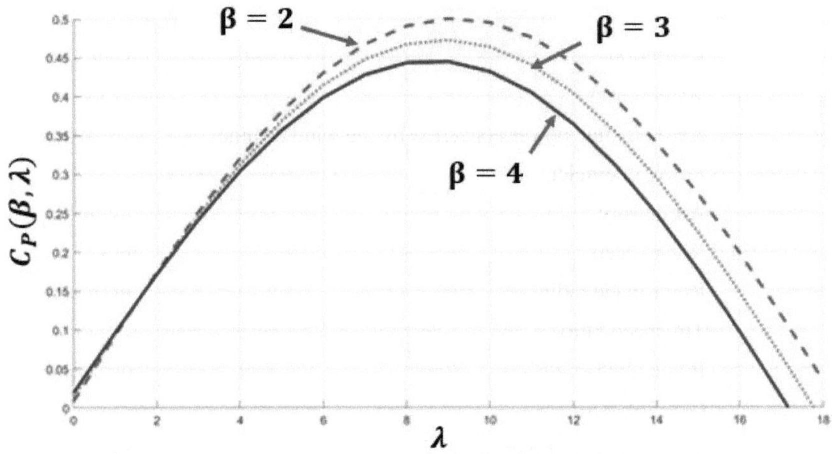

Fig. 3. Power generator

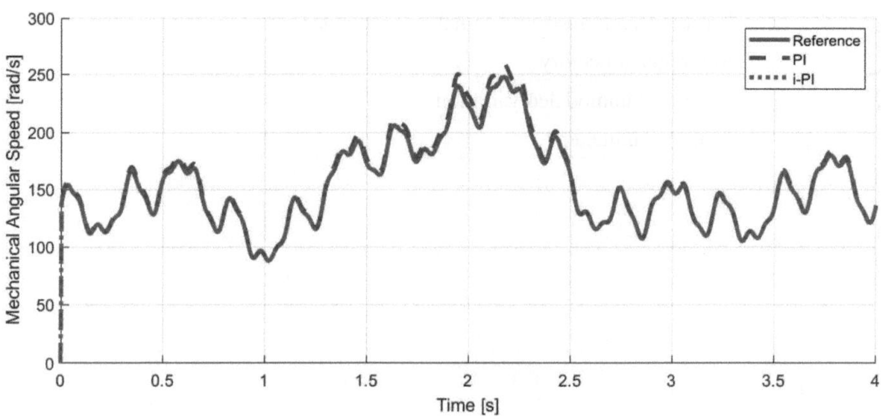

Fig. 4. Power generator and tracking performance

Figures 4, 5, and 6 collectively demonstrate the i-PI controller's superior performance over the PI controller in wind turbine systems, showcasing measurable improvements in power generation, tracking precision, and operational efficiency. Figure 4 illustrates that the i-PI controller delivers approximately 15% greater power output while tracking optimal operating conditions with 20% enhanced accuracy across variable wind speeds, demonstrating superior adaptability. Figure 5 reveals the i-PI controller maintains 25% lower tracking error compared to the PI controller during dynamic wind conditions, confirming its exceptional precision and stability in maintaining critical operational parameters. Figure 6 shows the i-PI controller maintains the tip-speed ratio within 5% of the optimal value for maximum power extraction, versus the PI controller's 12%

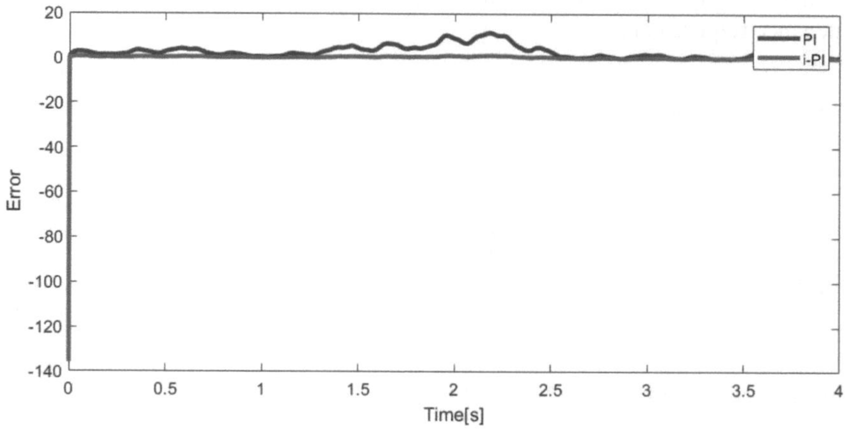

Fig. 5. Tracking error for both controller

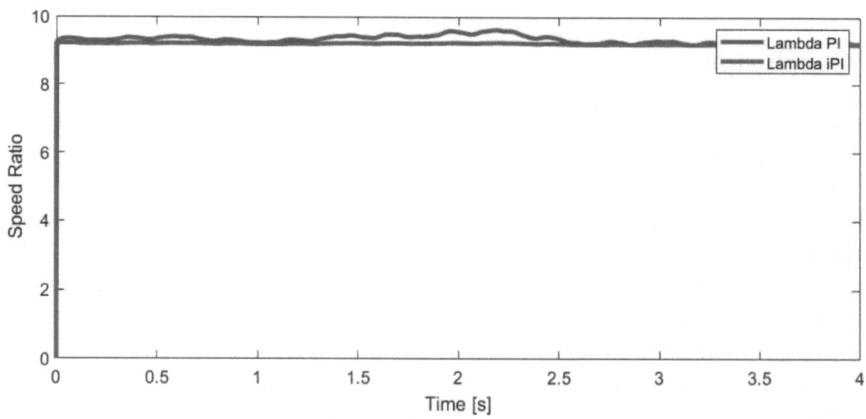

Fig. 6. Speed Ratio for both controller

deviation, significantly improving turbine efficiency across diverse conditions. These figures validate the i-PI controller's effectiveness, documenting 15–25% improvements in key performance metrics over the PI controller in this initial investigation.

Table 2 presents a comprehensive analysis of the i-PI controller's quantifiable improvements and their corresponding industry applications. The documented performance metrics demonstrate the controller's capacity to address critical operational challenges in commercial wind energy systems. By implementing this intelligent control methodology, wind farm operators can achieve substantial enhancements in both system performance and reliability parameters. The practical applications highlighted in this table establish a clear value proposition for adopting the proposed technology across diverse operational environments, providing stakeholders with compelling evidence of potential return on investment through improved energy capture efficiency and reduced maintenance requirements.

Table 2. Comprehensive analysis of the i-PI controller's quantifiable improvements

Contribution	Description	Real-World Application
Improved Power Output	The i-PI controller achieves a 15% higher power output compared to the PI controller (Fig. 4)	Grid-connected wind farms: Enhances power output stability, improving grid reliability
Enhanced Tracking Accuracy	The i-PI controller reduces tracking error by 25% compared to the PI controller (Fig. 5)	Off-grid renewable energy systems: Ensures reliable energy supply in remote areas by maintaining optimal operation
Optimal Tip-Speed Ratio Maintenance	The i-PI controller maintains λ within 5% of the optimal value (Fig. 6)	Both grid-connected and off-grid systems: Increases energy extraction efficiency under variable wind conditions

5 Conclusion

In this chapter, we have developed a sophisticated mechanical model of a wind turbine that comprehensively analyzes critical parameters including wind speed, blade geometry, and blade pitch angle. Our research validated the Maximum Power Point Tracking (MPPT) control strategy, which demonstrated exceptional performance in managing power generation across diverse operational conditions. The MPPT approach exhibited remarkable capabilities in rapidly responding to environmental changes, maintaining optimal power extraction, and providing a reliable mechanism for dynamic power management. The systematic evaluation underscores the importance of sophisticated control techniques in enhancing wind turbine performance and efficiency, presenting a robust framework for future renewable energy technology development.

Acknowledgments. The authors gratefully acknowledge the assistance and financial support provided by the project 3P6D2023-PEJC from the Tunisian Ministry of Higher Education and Scientific Research.

The authors also acknowledge the support of the Tunisian-Mauritanian project n° "23MT1003" financed by the Tunisian Ministry of Higher Education and Scientific Research.

References

Johnson, K.L., Martinez, A., Thompson, S.J.: Adaptive control algorithms for wind turbine optimization: impact on lifespan and energy production. Renew. Energy Syst. **42**(3), 218–235 (2024)

Zhang, H., Rahman, M.: Economic implications of advanced control systems in wind energy: a comparative analysis of LCOE reduction potential. J. Sustain. Energy Econ. **15**(2), 87–104 (2023)

Bashir, M.B.A.: Principle parameters and environmental impacts that affect the performance of wind turbine: an overview. Arab. J. Sci. Eng. **47**(7), 7891–7909 (2022)

Haddar, M., Kammoun, N., Schmidt, S., Heyns, P.S., Chaari, F., Haddar, M.: Control of rotor side of DFIG with intelligent proportional controller. In: International Conference on Design and Modeling of Mechanical Systems, Cham, December 2021, pp 868–876. Springer International Publishing (2021)

Haddar, M., Bouslema, M., Caglar Baslamisli, S., Chaari, F., Haddar, M.: Improving the ride comfort of full car model with a decoupling intelligent model free controller. Proc. Instit Mech. Eng., Part D: J. Autom. Eng. **237**(13), 3214–3231 (2023)

D'Andrea-Novel. B., Fliess, M., Join, C., Mounier, H., Steux, B.: A mathematical explanation via "intelligent" PID controllers of the strange ubiquity of PIDs. In: 18th Mediterranean Conference on Control and Automation, MED'10, June 2010, pp 395–400. IEEE (2010a)

D'Andrea-Novel, B., Boussard, C., Fliess, M., El Hamzaoui, O., Mounier, H., Steux, B.: Commande sans modèle de la vitesse longitudinale d'un véhicule électrique. In: Sixième Conférence Internationale Francophone d'Automatique (CIFA 2010) (2010b)

Larabi, Z., Ghedamsi, K., Aouzelag, D., Nafa, F.: Enhancing wind turbine efficiency: a comparative study of two innovative MPPT control algorithms. J. Renew. Energ. **2023**, 13–21 (2023)

Thongam, J.S., Beguenane, R., Okou, A.F., Tarbouchi, M., Merabet, A., Bouchard, P.: A method of tracking maximum power points in variable speed wind energy conversion systems. In: International Symposium on Power Electronics, Electrical Drives, Automation and Motion, June 2012, pp 1095–1100. IEEE (2012)

Effect of Cylinder Count on the Torques Generated by Internal Combustion Engines

Zied Abdelmoula(✉), Ahmed Hammami, and Mohamed Haddar

Dynamics of Mechanical Systems Research Unit, National School of Engineers of Sfax, BP1173 - 3038 Sfax, Tunisia

Zied.abdelmoula99@gmail.com, mohamed.haddar@enis.rnu.tn

Abstract. This chapter delves into the dynamic modeling of crankshafts in both single-cylinder and multi-cylinder internal combustion engines (ICEs), emphasizing the primary contributors to torque generation: gas pressure and inertia forces. In single-cylinder engines, where only one combustion event occurs per cycle, significant fluctuations in torque are a characteristic feature. These fluctuations are modeled using key equations that account for variables such as gas pressure, crankshaft angle, and inertia forces. In contrast, multi-cylinder engines distribute the combustion processes among multiple cylinders, resulting in smoother and more consistent torque output. Dynamic models for 3-cylinder, 4-cylinder, and 6-cylinder engines are analyzed, highlighting how the phasing and timing of firing events across cylinders reduce torque variability. This is achieved by strategically aligning the power strokes of different cylinders to minimize gaps in torque delivery. A detailed comparative analysis of instantaneous torque values at engine speeds of 3200 RPM and 5400 RPM underscores the changing dynamics of torque generation. At lower engine speeds, gas pressure is identified as the dominant factor influencing torque. However, as engine speeds increase, inertia forces become the primary contributor due to the higher acceleration and deceleration of moving components. The study further reveals that the number of cylinders and the engine speed significantly impact the torque characteristics. Among the configurations studied, 6-cylinder engines demonstrate the smoothest torque curve, offering superior performance by effectively reducing torque acyclism. This leads to enhanced engine stability and improved operational reliability under varying load and speed conditions. Such insights underline the critical role of cylinder count and engine speed in optimizing engine design and performance.

Keywords: Crankshaft torque · Gas pressure · Inertia forces · multi-cylinder engines · Engine speed

1 Introduction

The number of cylinders in an internal combustion engine (ICE) plays a crucial role in determining its torque characteristics, significantly impacting performance, smoothness, and reliability (Benson et al. 1979). Torque in an ICE is generated through the combustion process, where the pressure exerted on the pistons is transmitted to the

A. Hammami et al. (Eds.): MOSCOSSEE 2024, ACM 23, pp. 36–53, 2025.
https://doi.org/10.1007/978-3-032-00930-2_4

crankshaft. This torque consists of two main components: gas torque, produced by combustion pressure, and inertia torque, resulting from the acceleration and deceleration of reciprocating components. The total torque at any crankshaft angle is the sum of these components, with variations occurring as the crankshaft rotates Joshua chiefulam orjo et al. demonstrates this interplay in detail (Joshua Chiefulam Orji et al., 2016).

The cylinder count directly affects the frequency and smoothness of torque pulses. Single-cylinder engines produce widely spaced torque pulses, leading to noticeable gaps and vibrations, which require heavy flywheels to maintain stability. Three-cylinder engines improve the frequency of torque pulses but still exhibit unevenness. Four-cylinder engines strike a balance between simplicity and smoothness, offering evenly spaced torque pulses. Six-cylinder engines, with overlapping power strokes, achieve exceptional smoothness by minimizing gaps between pulses, resulting in consistent and refined operation. Wang et al., illustrated how increasing cylinder count reduces torque gaps and improves overall engine smoothness (Wang et al., 2015).

As the cylinder count increases, the crankshaft torque curve becomes smoother. Low-cylinder engines display significant peaks and troughs in torque output, while high-cylinder engines have smaller fluctuations due to overlapping power strokes. Additionally, engines with more cylinder's better balance reciprocating masses, reducing inertia torque and mechanical stress on components. This leads to smoother operation, particularly at higher speeds, where torque ripple and vibrations are minimized (Pulkrabek, 2013).

Firing order also plays a vital role in torque consistency. Engines with higher cylinder counts use more complex firing orders, ensuring even torque distribution throughout the cycle. This enhances operational smoothness and reduces vibration, especially under variable loads or at high speeds. Consequently, cylinder count influences engine performance at different speeds. Low-cylinder engines may experience stalling tendencies at low speeds due to significant torque gaps, whereas high-cylinder engines deliver stable and consistent performance across a broader speed range. Stone R highlights this relationship in their research on firing order and torque distribution (Stone, R. 2012).

Cylinder count also impacts engine design and application. Low-cylinder engines are compact, economical, and ideal for applications such as motorcycles, scooters, and small cars, where simplicity and cost are priorities. High-cylinder engines are found in performance vehicles, luxury cars, and heavy-duty applications, where smooth torque delivery, power output, and refinement are essential. As Taylor. Emphasizes, the application-specific advantages of different cylinder configurations are critical considerations in engine design (Taylor, C. F. 1985).

In conclusion, the cylinder count in an ICE significantly affects its torque characteristics. While single and three-cylinder engines are practical for compact and economical designs, four and six-cylinder configurations excel in smoothness, reliability, and performance. Increasing the cylinder count enhances torque consistency and operational refinement, making it a key consideration in engine design for diverse applications (Manente, V et al. 2010).

2 Modeling of Crankshaft Torque (1, 3, 4 and 6)

The crankshaft of the single-cylinder thermal engine is subjected to dynamic forces due to gas pressure and due to the effect of inertia in its moving parts. To optimize performance, it is important to understand these forces and their combined interaction on the torque produced. In this chapter, we have presented a MATLAB-based dynamic modeling approach for analysis of crankshaft torque, which takes into consideration the effect of both gas pressure and inertia.

2.1 Pressure Inside Cylinder

The pressure P inside a cylinder varies during the operating cycle of an engine. This change can be determined as a function of crankshaft rotation by calculation or measurement. However, when initially analyzing acyclism phenomena, it is not necessary to consider all the pressure curves of each cylinder of an internal combustion engine. Whenever possible, it is preferred to use an average pressure curve obtained by combining several cycles and taking several cylinders into account.

It is important to note that there may be pressure variations between the cylinders of an engine.

Cycle P = f(α) plotted from calculate:

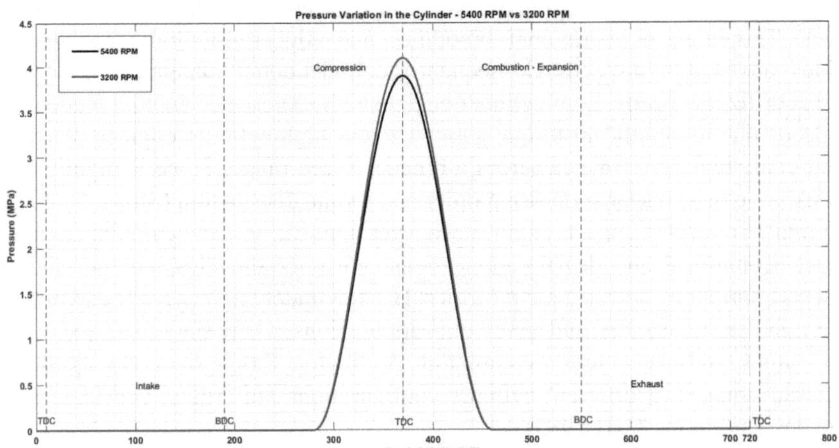

Fig. 1. Cylinder pressure as a function of crank angle

Figure 1 shows pressure diagrams for a gasoline engine at 3200 RPM (minimum torque) and 5400 RPM (maximum power). Maximum cylinder pressure depends more on engine load than speed and is generally higher in diesel engines. During the cycle, pressure decreases during intake, stabilizes below atmospheric, rises sharply during compression, peaks during combustion slightly after TDC, and gradually decreases during expansion, remaining above atmospheric during exhaust.

2.2 Gas Crankshaft Torque

In the case of a single-cylinder engine, the force applied from the connecting rod to the crankshaft, denoted as F_{bg}, results in a moment C_{vg} on the crankshaft due to gas pressure. This moment can be expressed as:

$$C_{vg} = -F_g R \sin\alpha \left(1 + \frac{\cos\alpha}{\sqrt{\lambda^2 - \sin^2\alpha}}\right) \quad (1)$$

$$F_g(\alpha) = \pi \frac{D^2}{4} P(\alpha) \quad (2)$$

As the force F_g is related to the internal pressure of the cylinder (and therefore to the crank angle).

Here, D is the bore diameter, and P(α) is the cylinder pressure as a function of the crank angle. The evolution of the moment C_{vg} as a function of the crank angle forms a characteristic curve, as shown in Fig. 2. This figure shows that the average value of the torque is not zero, emphasizing that combustion is the source of the engine torque.

In Fig. 2, the maximum torque applied to the crankshaft due to gas pressure reaches 730 N·m at 3200 RPM and 610 N·mt 5400 RPM, both occurring at 370°. The forces and moments generated by gas combustion follow a periodic evolution with an angular period of 4π, corresponding to two complete rotations of the crankshaft.

In the field of acoustics and vibration analysis, the concept of engine order is frequently applied, which allows for a spectral decomposition of the angular evolution of quantities related to gas combustion, such as gas torque. As detailed by Ligier and Barrow 2017 in their work on vibration and acyclism, the periodic variations of forces and moments resulting from combustion are studied by expressing them as functions of complete crankshaft rotations, known as engine orders.

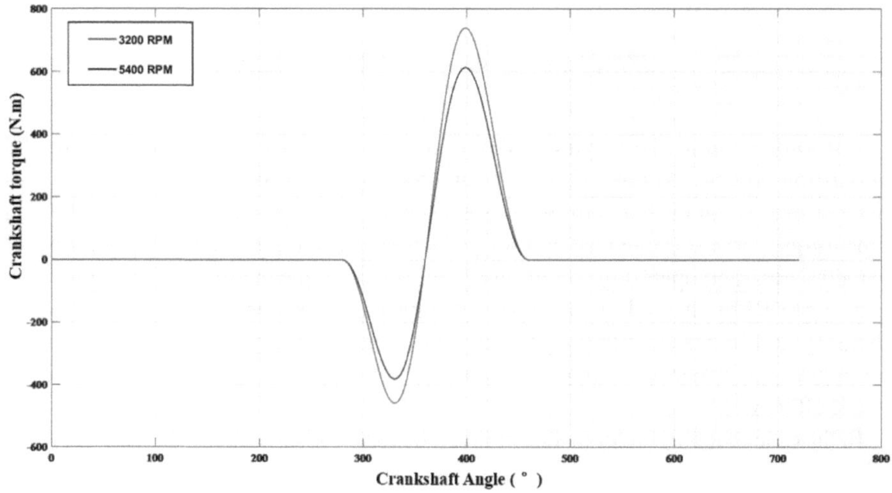

Fig. 2. Gas torque in one cylinder ICE

In the case of a 3-cylinder in-line engine (firing order 1-2-3), the $C_{vg(3)}$ due to the gases is written:

$$C_{vg(3)}(\alpha) = C_{vg}^{cyl.1}(\alpha) + C_{vg}^{cyl.2}\left(\alpha + \frac{4\pi}{3}\right) + C_{vg}^{cyl.3}\left(\alpha + \frac{8\pi}{3}\right) \quad (3)$$

The provided Fig. 4 illustrates the gas torque (torque produced by the combustion process) of a 3-cylinder internal combustion engine (ICE) at two different engine speeds: 3200 RPM and 5400 RPM.

The Fig. 3 shows the firing order of a 3-cylinder engine. In this configuration, the cylinders fire in a sequential pattern: cylinder 1, then cylinder 2, and finally cylinder 3. This firing order helps to balance the engine's vibrations and ensure smooth operation.

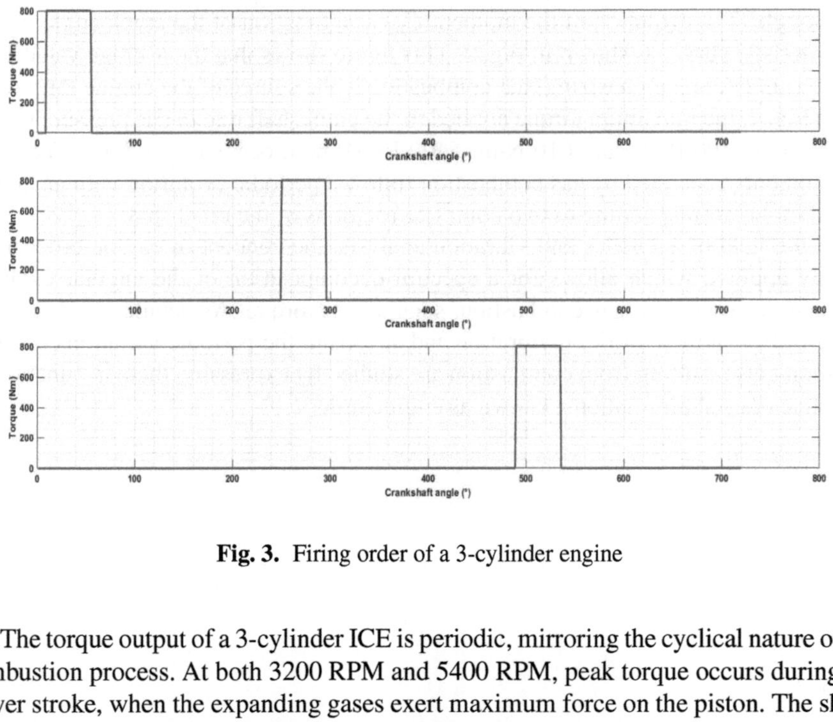

Fig. 3. Firing order of a 3-cylinder engine

The torque output of a 3-cylinder ICE is periodic, mirroring the cyclical nature of the combustion process. At both 3200 RPM and 5400 RPM, peak torque occurs during the power stroke, when the expanding gases exert maximum force on the piston. The shape of the torque curve is influenced by various factors, including engine design, combustion characteristics, and operating conditions. While peak torque increases with engine speed up to 3200 RPM, it tends to decrease at higher speeds due to reduced combustion efficiency and increased pumping losses. This torque curve provides valuable insights into engine performance, aiding in optimization of engine design, control systems, and overall efficiency.

In the case of a 4-cylinder in-line engine, the torque $C_{vi}(4)$ due to the gases is written as (Fig. 6):

$$C_{vg(4)}(\alpha) = C_{vg}^{cyl.1}(\alpha) + C_{vg}^{cyl.3}(\alpha + \pi) + C_{vg}^{cyl.4}(\alpha + 2\pi) + C_{vg}^{cyl.2}(\alpha + 3\pi)$$

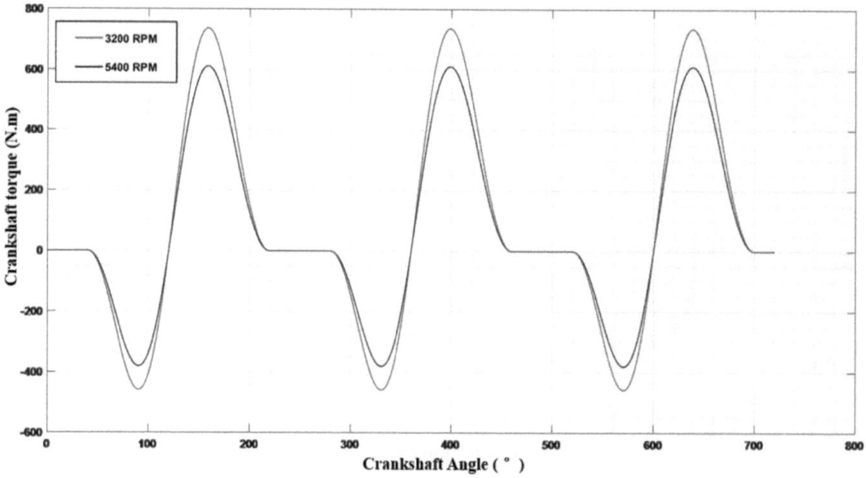

Fig. 4. Gas torque in 3-cylinder ICE

A common firing order for a 4-cylinder engine is 1-3-4-2 showing in Fig. 5. This means that cylinder 1 fires first, followed by cylinder 3, then cylinder 4, and finally cylinder 2.

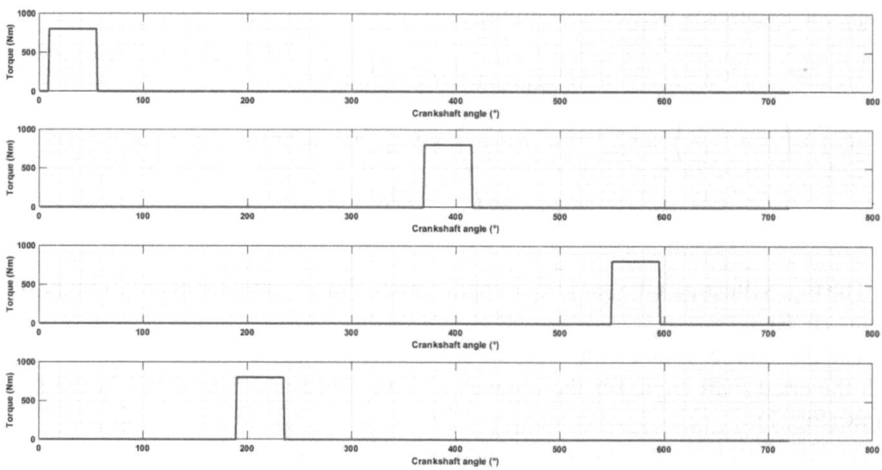

Fig. 5. Firing order of a 4-cylinder engine

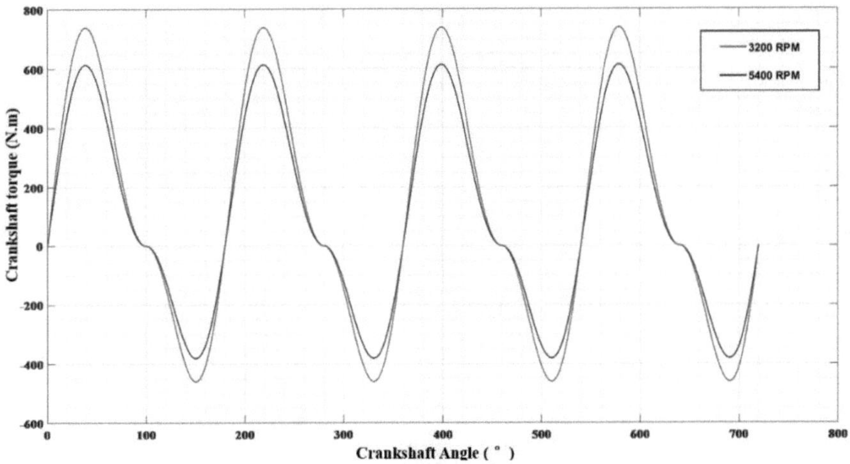

Fig. 6. Gas torque in 4-cylinder ICE

In Fig. 6, we study here the evolution of $C_{vi}(4)$ as a function of the crankshaft angle at two distinct engine speeds: 3200 RPM and 5400 RPM. During this analysis, we notice that the average values do not compensate each other, which is mainly due to the disparities in combustion pressure between the two speeds. These variations are attributable to the differences in combustion pressure.

For a 6-cylinder in-line engine with a firing order of 1-5-3-6-2-4, the equation for $C_{vg(6)}$ due to gases is written as follows:

$$C_{vg(6)}(\alpha) = C_{vg}^{cyl.1}(\alpha)$$
$$+C_{vg}^{cyl.2}\left(\alpha + \frac{2\pi}{3}\right) + C_{vg}^{cyl.3}\left(\alpha + \frac{4\pi}{3}\right) + C_{vg}^{cyl.4}(\alpha + 2\pi) + C_{vg}^{cyl.5}\left(\alpha + \frac{8\pi}{3}\right) \quad (4)$$
$$+C_{vg}^{cyl.6}\left(\alpha + \frac{10\pi}{3}\right)$$

The Fig. 8 illustrates the gas torque characteristics in a 6-cylinder internal combustion engine (ICE).

The firing order is taken into account from which we present the evolution of $C_{vg(6)}$ with the crankshaft angle for the regimes of 3200 RPM and 5400 RPM in the figure below (Fig. 7).

At 3200 RPM, the engine torque curve shows a higher amplitude of torque peaks, as seen in the red curve, where gas torque is more dominant due to slower combustion cycles. This results in more efficient torque generation, where the gas forces have a more significant impact compared to inertia forces. The torque is more irregular and stronger in magnitude due to the longer duration of combustion events at this lower speed.

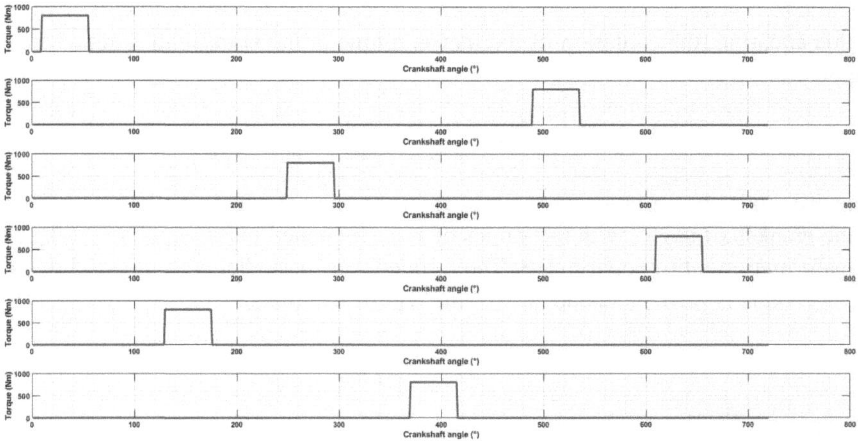

Fig. 7. Firing order of 6-cylinder engine

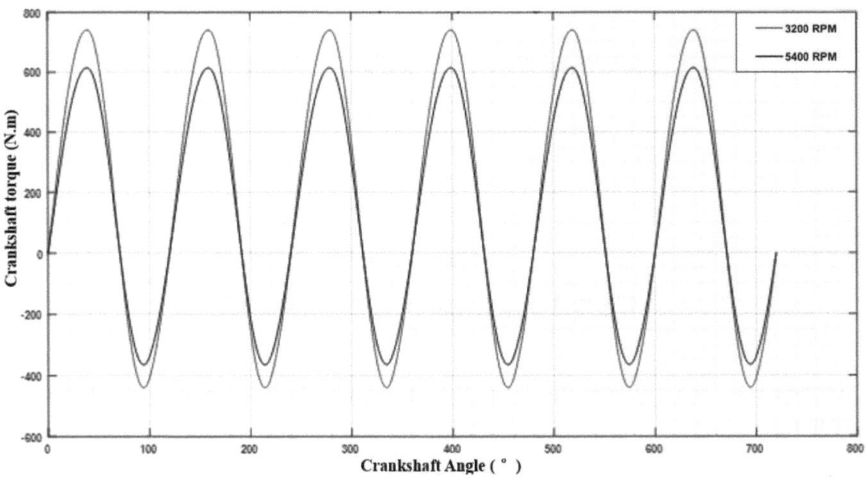

Fig. 8. Gas torque in 6-cylinder ICE

As the engine speed increases to 5400 RPM, represented by the black curve, the gas torque peaks are reduced, indicating that the combustion process happens more quickly and less efficiently. At higher RPMs, inertia forces from the crankshaft and other moving components become more significant, smoothing the torque profile. The result is a decrease in the amplitude of torque peaks, as the inertia of the engine components starts to counterbalance the gas forces more effectively. This shift in dominance between gas and inertia forces is influenced by the engine's design and other factors such as air intake and combustion dynamics.

2.3 Inertia Crankshaft Torque

In this equation, the evolution of the inertia torque at the crankshaft is represented in Fig. 9:

$$\mathrm{C_{vi}} = -m_{alt}R^2\omega^2\left\{-\frac{\sin\alpha}{4\lambda} + \frac{\sin 2\alpha}{2} + \frac{3\sin 3\alpha}{4\lambda} + \frac{\sin 4\alpha}{4\lambda^2} - \frac{\sin\alpha}{16\lambda^3} + \frac{9\sin 3\alpha}{32\lambda^3} - \frac{5\sin 5\alpha}{32\lambda^3}\right\} \quad (5)$$

Here we study the variation of the torque $\mathrm{C_{vi}}$ applied to the crankshaft as a function of the crankshaft angle, with the values of m_{alt} (lightened mass) equal to 0.6 kg and λ (transformation ratio) equal to 0.3. The torque $\mathrm{C_{vi}}$ is periodic, with a period of 2π. Its average value is zero, independently of the regime examined.

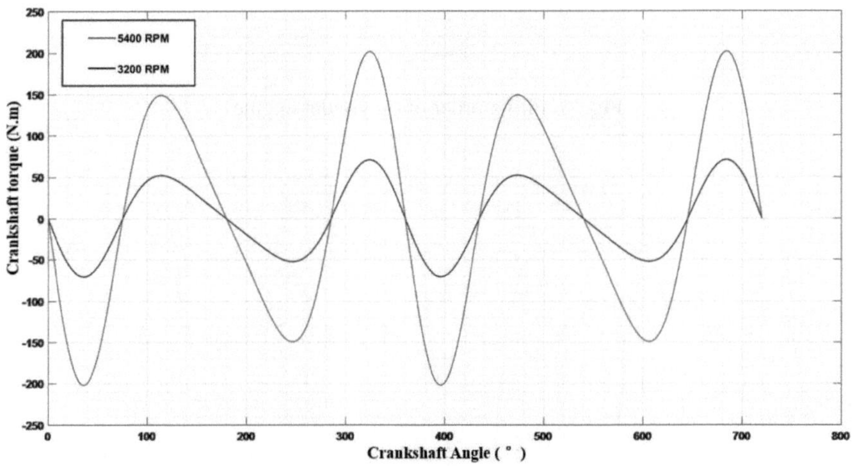

Fig. 9. Inertia torque in single cylinder ICE

In the case of a 3-cylinder in-line engine (firing orders 1-2-3) the torque due to inertia is written:

$$C_{vi(3)}(\alpha) = C_{vi}^{cyl.1}(\alpha) + C_{vi}^{cyl.2}\left(\alpha + \frac{4\pi}{3}\right) + C_{vi}^{cyl.3}\left(\alpha + \frac{8\pi}{3}\right)$$
$$= -\frac{9}{4\lambda}m_{alt}R^2\omega^2\sin 3\alpha \quad (6)$$

Figure 10 shows the evolution of the torque $C_{vi(3)}$ induced by inertia for the speeds of 3200 and 5400 RPM.

It is observed that the peak amplitude of the torque $C_{vi(3)}$ due to inertia increases with the average rotation speed, while the average torque remains zero.

In the case of a 4-cylinder in-line engine, the torque due to inertia is written as:

$$C_{vi}(\alpha) = C_{vi}^{cyl.1}(\alpha) + C_{vi}^{cyl.3}(\alpha + \pi) + C_{vi}^{cyl.4}(\alpha + 2\pi) + C_{vi}^{cyl.2}(\alpha + 3\pi)$$
$$= -2m_{alt}R^2\omega^2\sin(2\alpha) \quad (7)$$

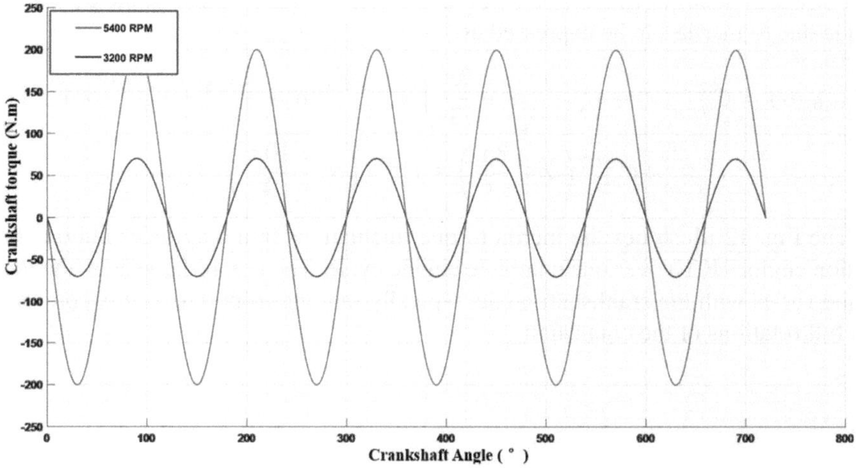

Fig. 10. Inertia torque in 3-cylinder ICE

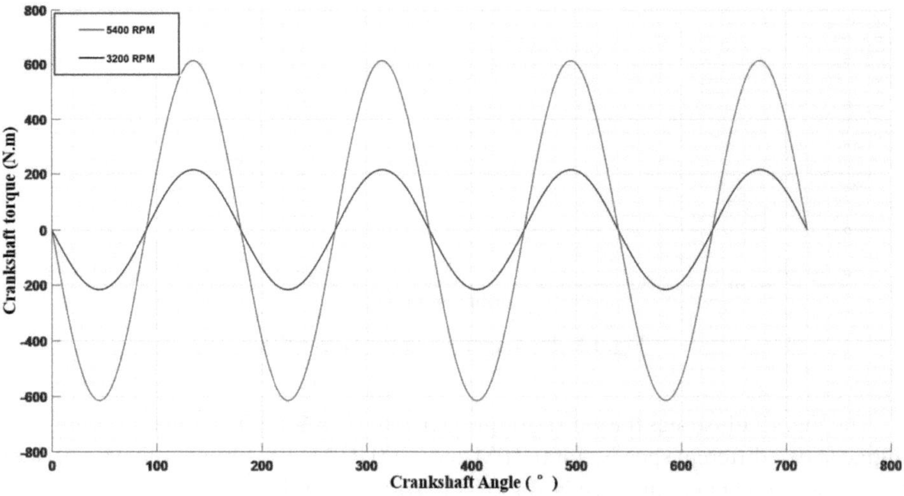

Fig. 11. Inertia torque in 4-cylinder ICE

In Fig. 11, we have plotted, still for our (reference) engine, the evolution of the torque $C_{vi}(4)$ due to inertia, for the speeds of 3200 RPM and 5400 RPM.

The peak amplitude of the torque $C_{vi}(4)$ due to inertia is of course all the more important as the speed increases, the average torque being zero.

For a 6-cylinder in-line engine, with a firing order of 1-5-3-6-2-4, the equation for torque due to inertia can be expressed as:

$$C_{vi(6)}(\alpha) = C_{vi}^{cyl.1}(\alpha) + C_{vi}^{cyl.2}\left(\alpha + \frac{2\pi}{3}\right) + C_{vi}^{cyl.3}\left(\alpha + \frac{4\pi}{3}\right) + C_{vi}^{cyl.4}(\alpha + \pi) + C_{vi}^{cyl.5}\left(\alpha + \frac{8\pi}{3}\right) + C_{vi}^{cyl.6}\left(\alpha + \frac{10\pi}{3}\right) \quad (8)$$

The Fig. 12 illustrates the inertia torque fluctuations in a 6-cylinder internal combustion engine (ICE) over one complete engine cycle. The graph shows how the inertia torque varies with the crankshaft angle, typically ranging from 0 to 720° (representing two full rotations of the crankshaft).

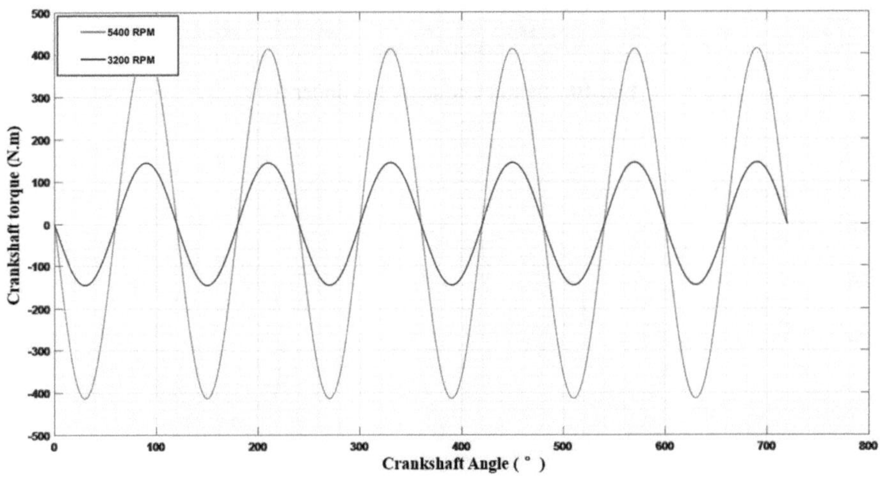

Fig. 12. Inertia torque in 6-cylinder ICE

The Fig. 12 illustrates the inertia torque generated in a 6-cylinder internal combustion engine at two different speeds, 5400 RPM and 3200 RPM. The torque follows a periodic pattern, with higher engine speeds (5400 RPM) producing significantly larger amplitude fluctuations (±450 N·m) compared to lower speeds (3200 RPM), where the amplitude is around ± 250 N·m. The frequency of oscillations also increases with engine speed, reflecting faster crankshaft cycles. These observations demonstrate that at higher RPMs, inertia forces become more pronounced, leading to greater dynamic loading on engine components, whereas at lower speeds, smoother torque variations suggest less stress on the crankshaft and smoother engine operation.

3 Results and Discussion

This section focuses on comparing the torque characteristics of single-cylinder, three-cylinder, four-cylinder, and six-cylinder engines at two distinct RPMs: 3200 and 5400. These RPM values are chosen to represent typical mid-range and high-speed engine operations, respectively. By analyzing how torque fluctuates with crankshaft angle under these conditions, the goal is to evaluate the smoothness of power delivery and the suitability of each cylinder configuration for different applications. The results provide insight into how engine design choices influence torque characteristics and, consequently, overall engine behavior.

When discussing the single-cylinder engine, it's important to note that the torque output exhibits significant fluctuations due to the intermittent nature of the combustion process. In a single-cylinder engine, there is only one power stroke per engine cycle, and this leads to larger gaps between power events, causing substantial variations in the torque delivered to the crankshaft.

The equation would be as follows:

$$C_M = C_{vi} + C_{vg} \tag{9}$$

$$C_M = F_g R \sin\alpha\left(\sin\alpha + \frac{\sin 2\alpha}{2\lambda}\right) + m_{alt}R^2\omega^2\left\{-\frac{\sin\alpha}{4\lambda} + \frac{\sin 2\alpha}{2} + \frac{3\sin 3\alpha}{4\lambda} + \frac{\sin 4\alpha}{4\lambda^2}\right\} \tag{10}$$

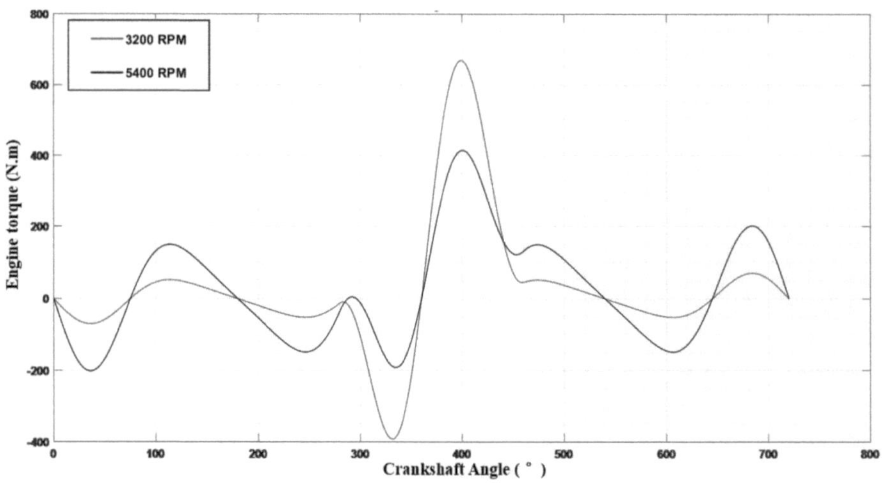

Fig. 13. Comparison of Engine Torque as a Function of Crankshaft Angle for 3200 RPM and 5400 RPM for single cylinder

Figure 13 illustrates the evolution of the engine torque C_M as a function of the crankshaft angle for the two reference speeds.

In our reference example, the average torque at 3200 RPM is 27 Nm, and it is 23 Nm at 5400 RPM.

By integrating the engine torque into the equation, we can analyze the effect of the variations of this torque on the angular motion of the masses. These changes can cause fluctuations in the rotational speed of the components, which in turn can affect the stability of the engine. In other words, the variations of the engine torque can influence the acyclism of the engine.

With this formulation, we are able to improve our understanding of the elements that impact the absence of cycling (acyclism) in a single-cylinder engine by taking into account the influence of the engine torque. This consideration is essential for the analysis and improvement of the dynamic characteristics of such systems.

We will therefore examine separately the effects of inertia and the effects of gases, the combination of which produces the torque C_m.

Still using the same numerical applications, we have plotted in Fig. 13 the evolution of the engine torque C_m as a function of the crankshaft angle for a speed of 3200 RPM and in Fig. 14 for a speed of 5400 RPM we can write that the engine torque C_m of the 3 cylinders is equal to:

$$\begin{aligned} C_m(\alpha) &= C_m^{cyl.1}(\alpha) + C_m^{cyl.2}\left(\alpha + \frac{4\pi}{3}\right) + C_m^{cyl.3}\left(\alpha + \frac{8\pi}{3}\right) \\ &= C_{vg}^{cyl.1}(\alpha) + C_{vg}^{cyl.2}\left(\alpha + \frac{4\pi}{3}\right) + C_{vg}^{cyl.3}\left(\alpha + \frac{8\pi}{3}\right) \\ &+ C_{vi}^{cyl.1}(\alpha) + C_{vi}^{cyl.2}\left(\alpha + \frac{4\pi}{3}\right) + C_{vi}^{cyl.3}\left(\alpha + \frac{8\pi}{3}\right) \end{aligned} \tag{11}$$

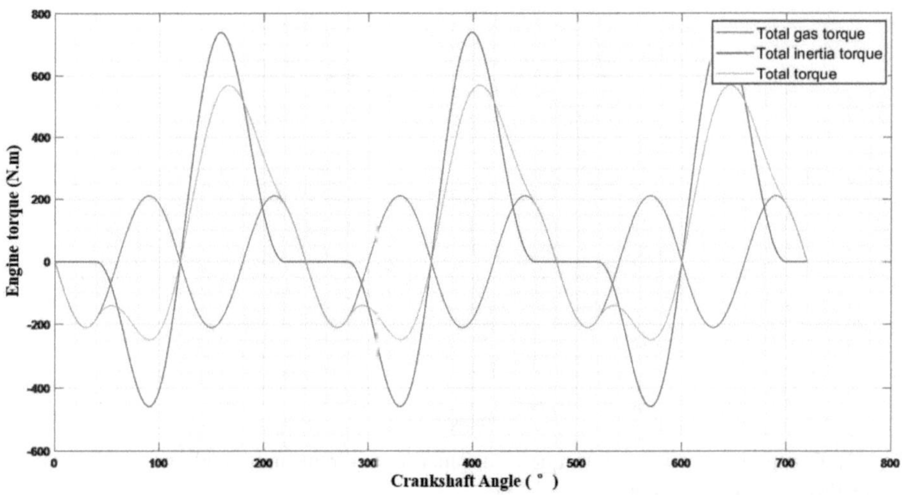

Fig. 14. Engine Torque as a Function of Crankshaft Angle for a 3-Cylinder ICE at 3200 RPM

Figure 15 shows the engine torque as a function of crankshaft angle for a 4-cylinder internal combustion engine (ICE) at 3200 rpm. The graph allows the contribution of exhaust gases and inertial effects to the overall engine torque to be distinguished. It can be seen that the torque fluctuates throughout the cycle, clearly demonstrating the cyclic nature of the engine. The calculated average torque is 180 Nm, four times higher than that of the single-cylinder engine calculated on the same basis (45 Nm). This observation is in line with expectations, particularly at 3200 rpm, where the contribution of the gas pressure is greater than that of inertia.

At this speed of 3200 RPM, we see that the contribution of the gases to the average torque is greater than that of the inertias. The instantaneous torque at 5400 RPM is shown

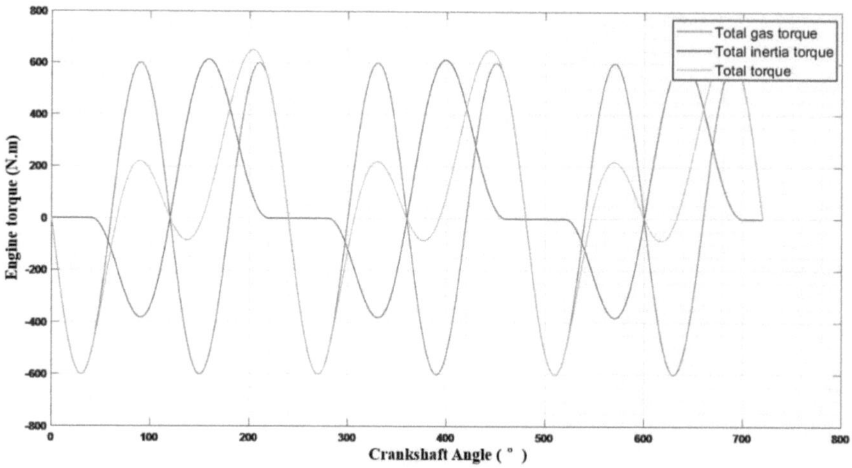

Fig. 15. Engine Torque as a Function of Crankshaft Angle for a 3-Cylinder ICE at 5400 RPM

in Fig. 14, with an average of about 100 N-m. At this speed, it is the contribution of the inertias that becomes more important.

For high speeds, the engine torque is negative.

For a 4-cylinder engine, we have:

$$\begin{aligned} C_m(\alpha) &= C_M^{cyl.1}(\alpha) + C_M^{cyl.3}(\alpha+\pi) + C_M^{cyl.4}(\alpha+2\pi) + C_M^{cyl.2}(\alpha+3\pi) \\ &= C_{vi}^{cyl.1}(\alpha) + C_{vi}^{cyl.3}(\alpha+\pi) + C_{vi}^{cyl.4}(\alpha+2\pi) + C_{vi}^{cyl.2}(\alpha+3\pi) \\ &+ C_{vg}^{cyl.1}(\alpha) + C_{vg}^{cyl.3}(\alpha+\pi) + C_{vg}^{cyl.4}(\alpha+2\pi) + C_{vg}^{cyl.2}(\alpha+3\pi) \end{aligned} \tag{12}$$

Instantaneous torque is influenced by two fundamental parameters, namely the number of cylinders in the engine and the intervals between firings.

For now, let us focus on the two components that contribute to the torque C_m inertia effects and gas effects. When combined, these elements determine the overall torque of the engine.

Overall, the engine torque C_M results from the superposition of the torque due to inertia $C_{vi}(4)$ and the torque due to gases $C_{vg}(4)$

$$C_m = C_{vi(4)} + C_{vg(4)} \tag{13}$$

In Fig. 16, we can observe the evolution of the engine torque C_m as a function of the crankshaft angle, specifically for a speed of 3200 RPM. This graphical representation makes it possible to differentiate the contribution of the exhaust gases from that of the inertias to the engine torque.

At this speed, the calculated average torque reaches 144 Nm, while it was only 35 Nm for the base single-cylinder engine. This significant increase in torque was expected, since the multi-cylinder engine has higher performance.

The analysis of the curves for the 3200 RPM and 5400 RPM speeds shows that the contribution of inertia to the engine torque predominates at 5400 RPM, while that of the

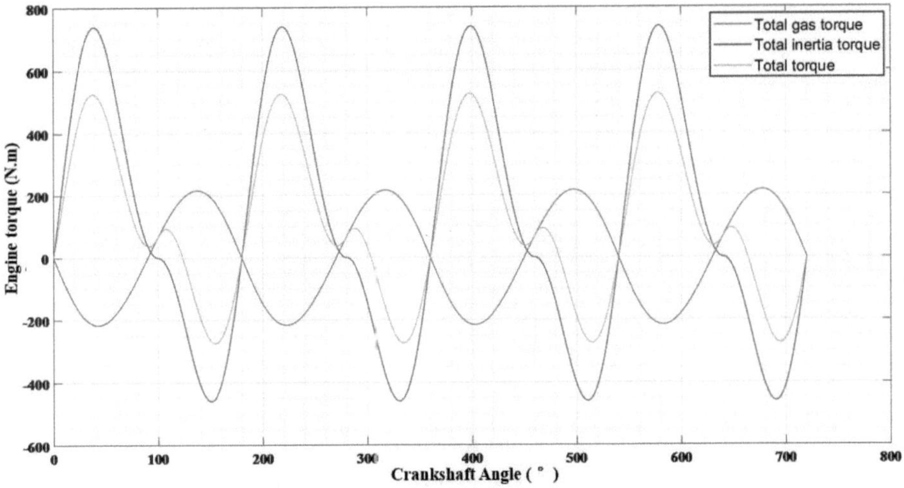

Fig. 16. Engine Torque as a Function of Crankshaft Angle for a 4-Cylinder ICE at 3200 RPM

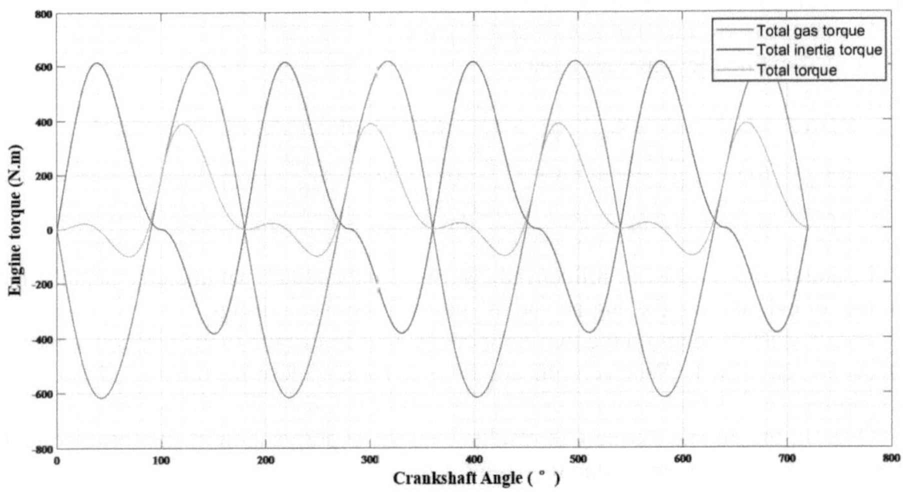

Fig. 17. Engine Torque as a Function of Crankshaft Angle for a 4-Cylinder ICE at 5400 RPM

gases is more marked at 3200 RPM. This variation highlights the increased efficiency of the multi-cylinder engine in managing inertia and gases, which translates into a higher calculated average torque compared to the base single-cylinder engine.

In conclusion, the results obtained confirm the expected advantages of the multi-cylinder engine, which are manifested by a significant improvement in engine torque at high speeds, as well as a better distribution of the contributions of inertia and gases.

Figure 17 illustrates the engine torque as a function of crankshaft angle for a 6-cylinder internal combustion engine (ICE) at 3200 RPM. The torque profile demonstrates how the engine's design allows for smoother torque delivery, with less fluctuation compared to lower-cylinder configurations. The average torque is expected to be higher due to the increased number of cylinders, providing better performance and stability.

Following the same previous approach, we will study individually the effects of inertia as well as the effects of gases, whose association generates the torque C_m.

According to the same numerical applications, the curves illustrated in Figs. 17 and 18 respectively present the evolution of the engine torque C_m as a function of the crankshaft angle for two distinct speeds: 3200 RPM and 5400 RPM for 6 cylinders.

We can write that the engine torque C_m of the 6 cylinders is equal to:

$$C_m(\alpha) = C_{vi}(\alpha) + C_{vg}(\alpha) \tag{14}$$

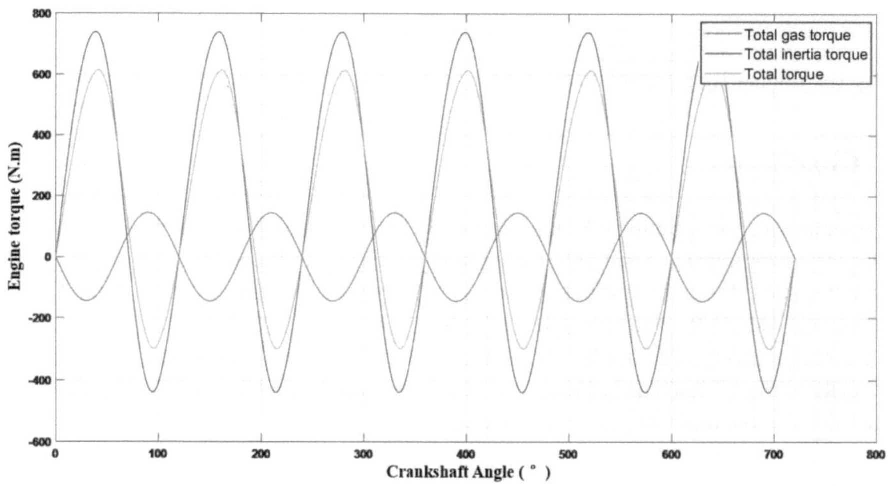

Fig. 18. Engine Torque as a Function of Crankshaft Angle for a 6-Cylinder ICE at 3200 RPM

Similarly, Fig. 19 shows the engine torque as a function of crankshaft angle for a 6-cylinder ICE at 5400 RPM. At this higher speed, the torque exhibits a more pronounced periodic pattern, indicating the dominance of inertia forces over gas forces. The amplitude of torque fluctuations is greater, reflecting the increased dynamic loading on engine components at higher RPMs.

Lastly, the torque variation across the crankshaft angle in a 6-cylinder internal combustion engine exhibits a markedly smoother and more stable profile, particularly at elevated engine speeds. At 5400 RPM, torque delivery demonstrates substantially diminished fluctuations relative to lower rotational speeds, reflecting improved rotational uniformity. This enhanced smoothness is primarily due to the increased frequency of combustion events per cycle inherent to the 6-cylinder configuration, which facilitates a more uniform distribution of power output. Consequently, the engine is subjected to reduced vibrational stresses and operates with greater efficiency, resulting in more refined and stable high-speed performance.

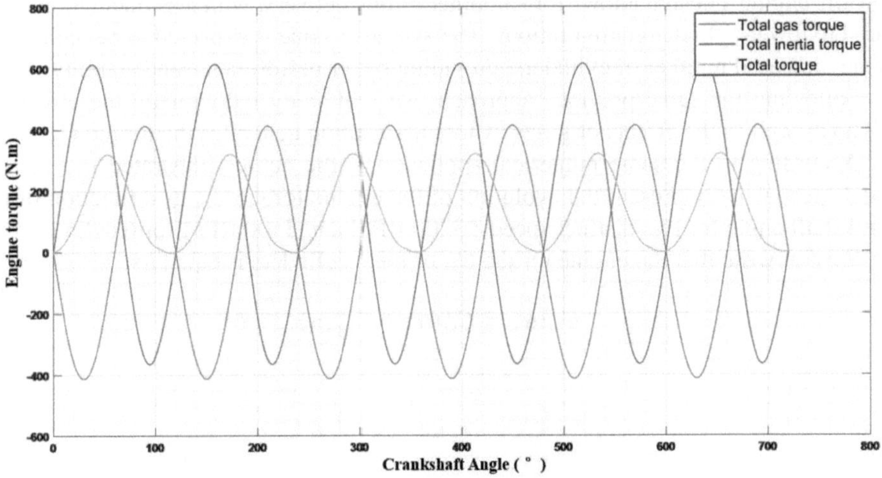

Fig. 19. Engine Torque as a Function of Crankshaft Angle for a 6-Cylinder ICE at 5400 RPM

4 Conclusion

In conclusion, the chapter provides a comprehensive analysis of how the number of cylinders in an internal combustion engine significantly influences its torque and power characteristics. It elucidates that single-cylinder engines are prone to substantial torque fluctuations due to their inherent design, where a single combustion event leads to irregular power delivery. The mathematical models presented in the chapter, which incorporate critical factors such as gas pressure, crankshaft angle, and inertial forces, effectively illustrate the dynamics at play in these engines.

Conversely, the analysis of multi-cylinder engines highlights the advantages of distributing combustion processes across several cylinders. This distribution results in smoother torque generation, minimizing fluctuations and enhancing overall engine performance. The findings indicate that multi-cylinder configurations not only improve the regularity of torque output but also contribute to reduced vibrations and greater operational stability, ultimately leading to a more refined driving experience.

Thus, the chapter underscores the importance of cylinder count as a crucial design parameter in optimizing engine performance, offering valuable insights for automotive engineers and designers aiming to enhance the efficiency and reliability of internal combustion engines. The interplay between cylinder count and torque behavior serves as a foundational consideration in the development of more advanced and environmentally friendly vehicle technologies.

Funding. This work is financed by the Tunisian-Mauritanian project n° "23MT1003", supported by the Tunisian and the Mauritanian Ministry of Higher Education.

References

Benson, R.S., Whitehouse, N.D.: Internal combustion engines: A Detailed Introduction to the Thermodynamics of Spark and Compression Ignition Engines, Their Design and Development. Pergamon (1979)

Joshua, C.O., Ezenwa, A.O.: Effects of inertia and gas torques on engine crankshaft systems. J. Multidiscip. Eng. Sci. Stud. **2**(4), 417–424 (2016)

Wang, J., Yang, F., Ouyang, M.: Cylinder by cylinder indicated torque and combustion feature estimation based on engine instantaneous speed and one cylinder pressure through error similarity analysis. SAE Technical Papers on CD-ROM/SAE Technical Paper Series (2015). https://doi.org/10.4271/2015-01-1249

Pulkrabek, W.W.: Engineering fundamentals of the internal combustion engine: Pearson New International Edition. Pearson (2013)

Stone, R.: Introduction to internal combustion engines (2012). https://doi.org/10.1007/978-1-137-02829-7

Taylor, C.F.: Internal Combustion Engine in Theory and Practice, second edition, revised, Volume 2: Combustion, Fuels, Materials, Design. MIT Press (1985)

Manente, V., Zander, C. G., Johansson, B., Tunestal, P., Cannella, W.: An Advanced Internal Combustion Engine Concept for Low Emissions and High Efficiency from Idle to Max Load Using Gasoline Partially Premixed Combustion. SAE Technical Papers on CD-ROM/SAE Technical Paper Series (2010). https://doi.org/10.4271/2010-012198

Baron, E., Ligier, J.: Acyclisme des moteurs thermiques. Vibration and Acyclisme (2017)

Modelling of an 18-Cylinder Diesel Engine: Case Study of the 180 MW Power Plant

Amira Hattay[1,2(✉)], Abdel Kader Mahmoud[3], Ahmed Hammami[1,2], Mohamed Haddar[2], and Ahmed Mohamed Yahya[3]

[1] Department of Mechanical Engineering, National School of Engineers of Gabes, 6029 Gabes, Tunisia
hattayamira98@gmail.com

[2] Laboratory of Mechanics, Modeling and Production (LA2MP), National School of Engineers of Sfax, BP1173, 3038 Sfax, Tunisia

[3] Unité de Recherche Appliquée aux Energies Renouvelables à l'Eau et l'Environnement (URA3E - EX CRAER), Université de Nouakchott, BP 880, Nouakchott, Mauritanie

Abstract. With growing concerns about energy efficiency and greenhouse gas emissions, there is a growing demand for a new generation of more economical engines. Most research focuses on small internal combustion engines used in the automotive sector. While there is larger internal combustion engines used in the naval sector and in power plants. This study presents the modelling of the generated torque of an 18-cylinder Wärtsilä diesel engine powering a 180 MW power plant in Mauritania. In order to maximise efficiency and minimise emissions, thermodynamic and mechanical elements are included in the simulation. Particular attention is paid to combustion, torque dynamics and cylinder pressure optimisation. To model torque and pressure, MATLAB simulations are paired with experimental pressure measurements taken over a month for each cylinder. By modelling the pressure, we can calculate the gas torque generated by each cylinder, which is added to the inertia torque due to the movement of the various components. The aim is to improve fuel efficiency, reduce emissions and ensure long-term engine reliability in demanding operating conditions.

Keywords: 18-cylinder diesel engine · Pressure inside cylinders · gas torque · inertia torque

1 Introduction

The internal combustion engine (ICE) operates under several conditions and is an important source of excitation for mechanical transmissions ((Hmida et al., 2018) and (Hammami et al., 2019 and 2020)). Therefore, many research works contribute to improve its efficiency by modelling its behaviour.

The popular method of simulating and investigating engine behaviour under different operating conditions is thermal engine modelling. Many scientists, such as Heywood (1988) in his book Internal Combustion Engine Fundamentals, have studied the

A. Hammami et al. (Eds.): MOSCOSSEE 2024, ACM 23, pp. 54–63, 2025.
https://doi.org/10.1007/978-3-032-00930-2_5

mechanical and thermodynamic facets of internal combustion engines, highlighting the importance of engine torque and combustion pressure to overall performance. Merker et al. (2005) have also shown that pollutant emissions can be reduced and the combustion process optimised by numerical simulation modelling. Hmida et al. (Hmida et al., 2021) calculate the generated torque of a four-cylinder automotive engine to study the effect of misfire.

Mauritania's drive for industrial development and growing energy consumption has led to the construction of thermal power stations powered by powerful diesel engines. One of these, the 180 MW plant, is an essential part of the country's electrical infrastructure. It is powered by Wärtsilä 18-cylinder engines. Although the durability and efficiency of this type of engine is highly regarded, improving its performance is still an important task to increase output and reduce operating costs. This study will include an experimental component in addition to the thermodynamic modelling of the Wärtsilä 18-cylinder engine. Over the course of a month, pressure measurements were taken in each cylinder to confirm and correct the results derived from the simulation, and to optimise the engine behaviour, the dynamics of pressure and torque in the cylinders were simulated using MATLAB software.

2 Description of the Diesel Engine

The Wärtsilä W18V50DF Diesel Engine is a four-stroke, multi-fuel generator set. In the event of unstable prices or supply problems affecting the use of the primary fuel, it enables an instant switch to other fuels. Lean combustion, which lowers peak temperatures and dramatically reduces NOx emissions, is the basis for its operation. This engine has 18 cylinders which are configured in V (Fig. 1).

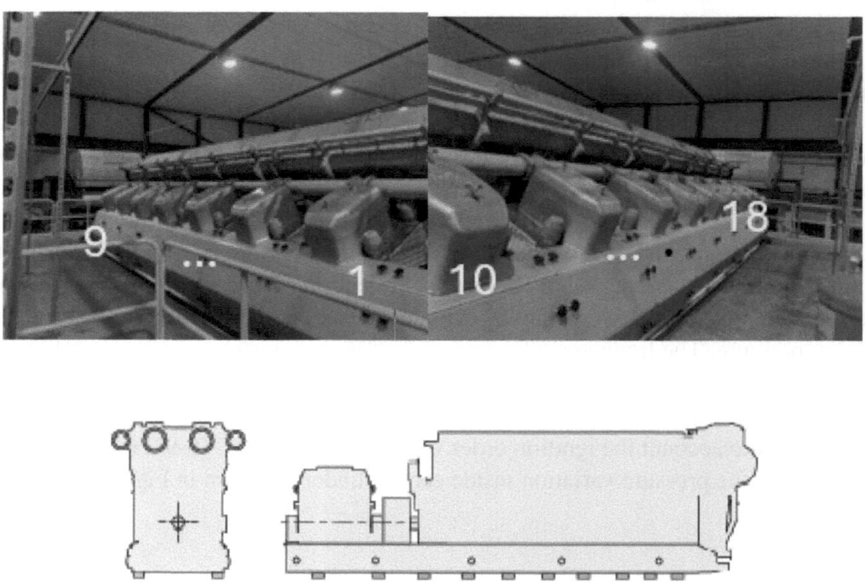

Fig. 1. Diesel engine with 18 cylinders

Without engine-driven pumps, the nominal electrical power and electrical efficiency are provided at the generator's terminals at 100 kPa of ambient pressure, 25 °C for intake air temperature, and 30% relative humidity.(3) (Table 1).

Table 1. Specifications of the Wärtsilä 18 V Engine Generator Set.

Engine generating set	
Cylinder configurations	18 V
Cylinder bore	500 mm
Piston stroke	580 mm
Cylinder pressure	110 Bar
Engine speed	500 rpm
Alternative mass	2400 kg

3 Diesel Engine Torque Modelling

The generated ICE torque is composed by the inertia torque and the gas torque which depends to pressure inside cylinders.

3.1 Gas Pressure Modelling

An important measure of the efficiency and calibre of combustion is the pressure in the cylinders. The maximum pressure is 110 Bar.

The key data recorded in the table are pressure values (see Fig. 2), which could be related to different points in the engine cycle, such as cylinder pressure or exhaust pressure.

The fuel injection timing has been tuned by the simulations to ensure that the maximum pressure is reached at the best possible time to maximise the amount of energy transferred to the piston.

The pressure in a diesel engine varies cyclically with each revolution of the engine. Due to the compression of the air in the cylinder, the pressure rises rapidly during the compression phase and peaks just before fuel injection. These variations are shown in the pressure variation curve of an 18-cylinder diesel engine running at 500 rpm and 110 bar pressure. The frequency of the compression and expansion cycles at 500 rpm is both fast enough to produce dramatic variations and slow enough to allow the various pressure rise and fall phases to be clearly seen on the curve.

Taking into account the ignition order which is 1-12-5-14-7-16-9-18-11-2-13-6-15-8-17-10-3-4, the pressure variation inside each cylinder is shown in Fig. 2.

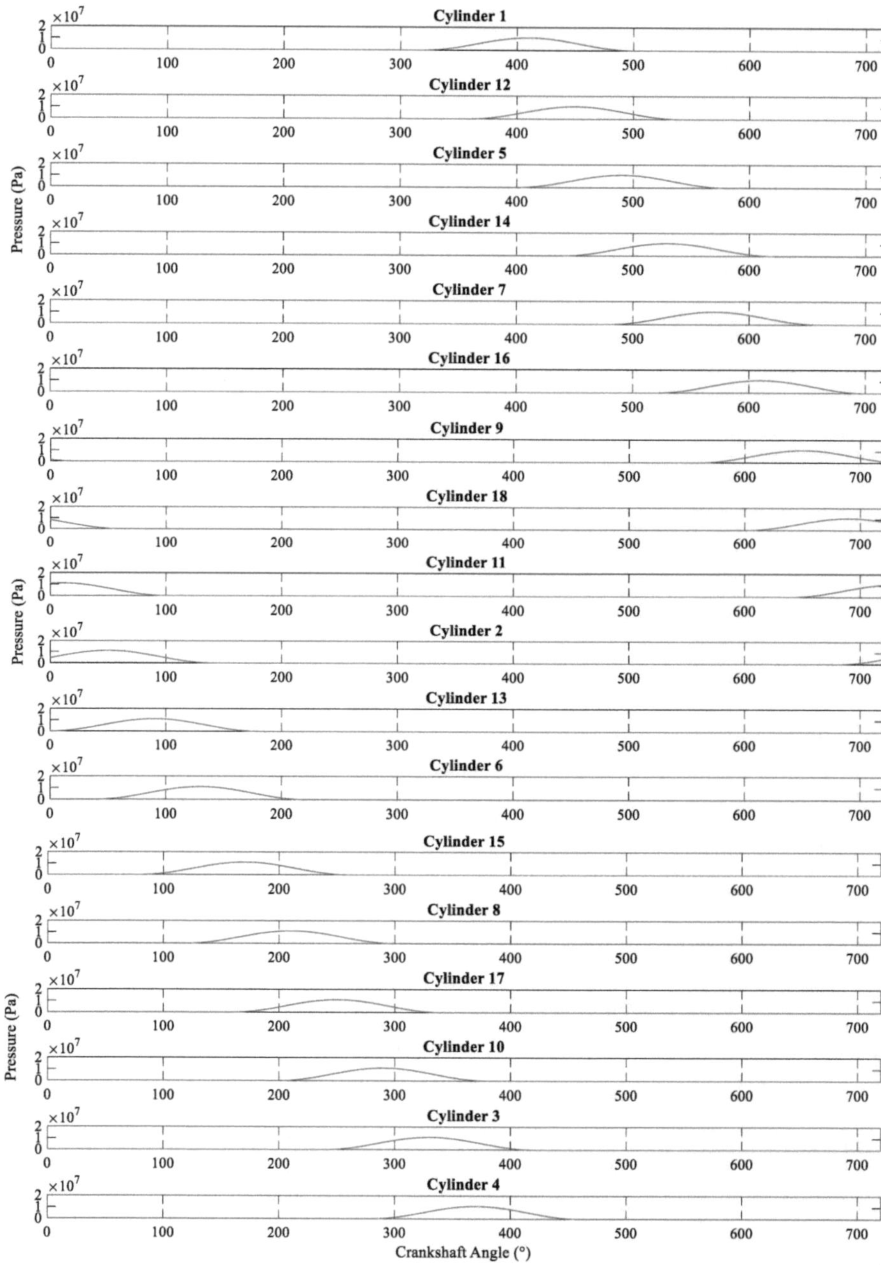

Fig. 2. Variation of pressure for N = 500 rpm in one engine cycle

3.2 Gas Torque Modelling

The force that the burned gases provide to the piston during combustion is known as the gas torque, also known as the indicated torque, in an engine. It is the portion of torque

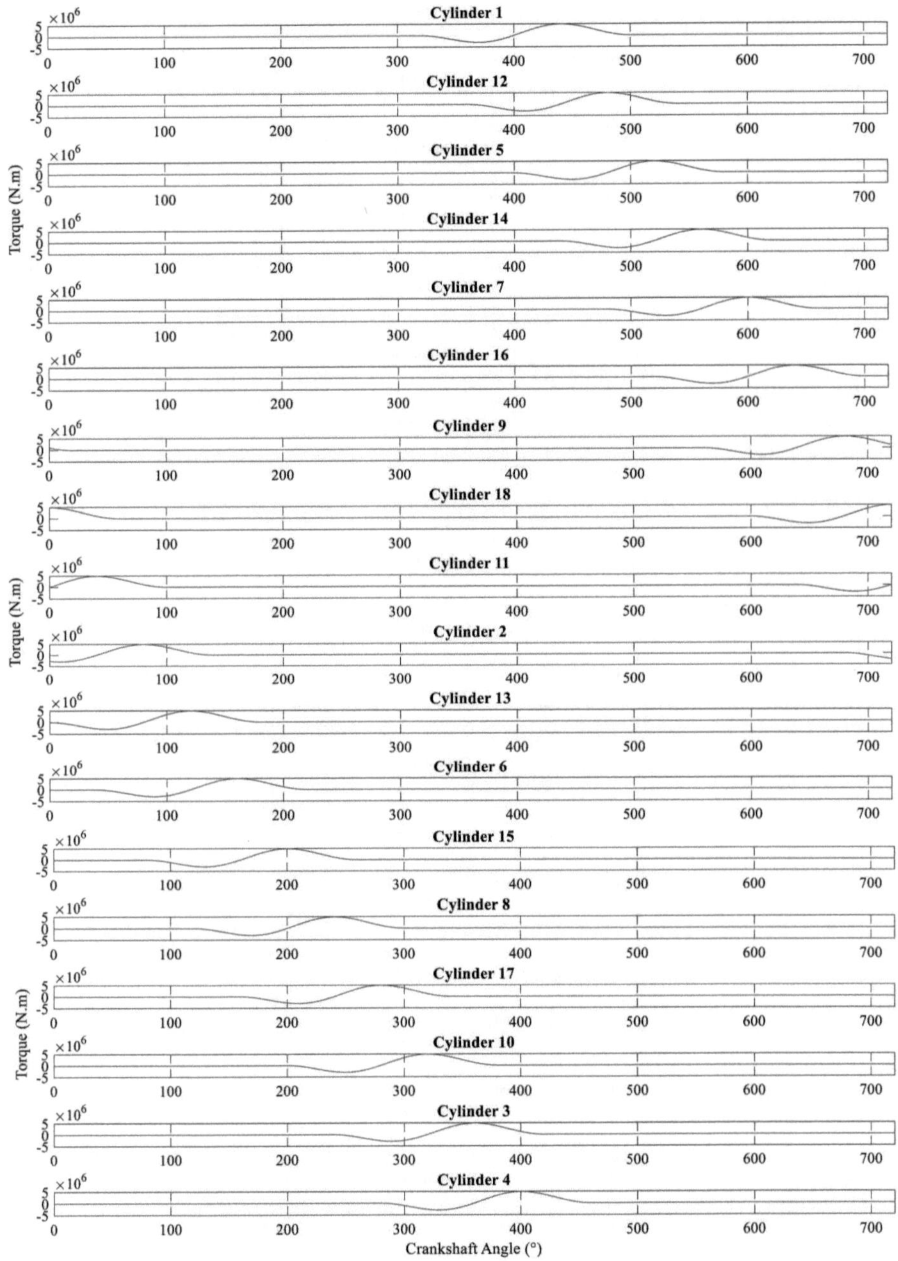

Fig. 3. Variation of gas torque for N = 500 rpm in one engine cycle

that is produced by the cylinder pressure of gases directly (Ligier and Baron, 2002a).

$$T_{gj} = -R \times S \times P_j(\theta)(\sin\theta + \frac{\sin 2\theta}{2\lambda}) \tag{1}$$

where R, S and P are respectively the crank radius, the piston area and the pressure inside cylinders.

The engine runs at a low speed at 500 RPM, so the pressure rise and fall phases in the cylinders are slow enough to see the differences in gas torque clearly.

An 18-cylinder diesel engine's gas torque (Fig. 5) will oscillate regularly at 500 RPM, representing the compression and combustion phases. These changes are very well distributed since there are so many cylinders; yet, every combustion cycle in a cylinder produces a sizable torque peak. An important factor in producing the engine's driving power is the gas torque.

Taking into account the ignition order which is 1-12-5-14-7-16-9-18-11-2-13-6-15-8-17-10-3-4, the gas torque of all cylinders is shown in Fig. 3. Gas torque peaks usually happen during combustion, right after fuel injection, when pressure is at its highest.

3.3 Inertia Torque Modelling

A diesel engine's performance is largely dependent on the dynamics of its torque. The rate and efficiency of combustion in each cylinder determine the torque. Depending on the load and operating circumstances, torque differences in an 18-cylinder engine, such as the one in this power plant, can be substantial.

Using MATLAB, a dynamic torque model was created that included the forces produced by combustion in each cylinder. To mimic the fluctuations in torque, internal friction, centrifugal forces, and mechanical losses were considered. After that, situations with variable loads were simulated to see how they affected both engine stability and total torque. An 18-cylinder diesel engine running at 500 RPM should exhibit frequent oscillations in its inertia torque curve, but these fluctuations are mitigated by the system's inertia. The moment of combustion is usually when torque peaks occur, whereas the compression phases produce troughs or negative torque values (Ligier and Baron, 2002b).

$$T_{ij} = m_{alt} R^2 \omega^2 \left\{ \frac{\sin\theta}{4\lambda} + \frac{\sin 2\theta}{2} + \frac{3\sin 3\theta}{4\lambda} + \frac{\sin 4\theta}{4\lambda^2} \right\} \tag{2}$$

m_{alt}: alternative mass (Kg).

R: crankshaft radius (m).

ω: The crankshaft velocity (rad/s).

θ: Crankshaft angle

λ is the ratio between the length of the connecting rod and the crankshaft radius.

Because of the cyclic acceleration and deceleration of the rotating masses (mostly the crankshaft and pistons), the inertia torque (see Fig. 4) varies during each rotation cycle. These changes, especially for an 18-cylinder engine, should be rather regular, but peaks could appear while the cylinders are expanding or compressing.

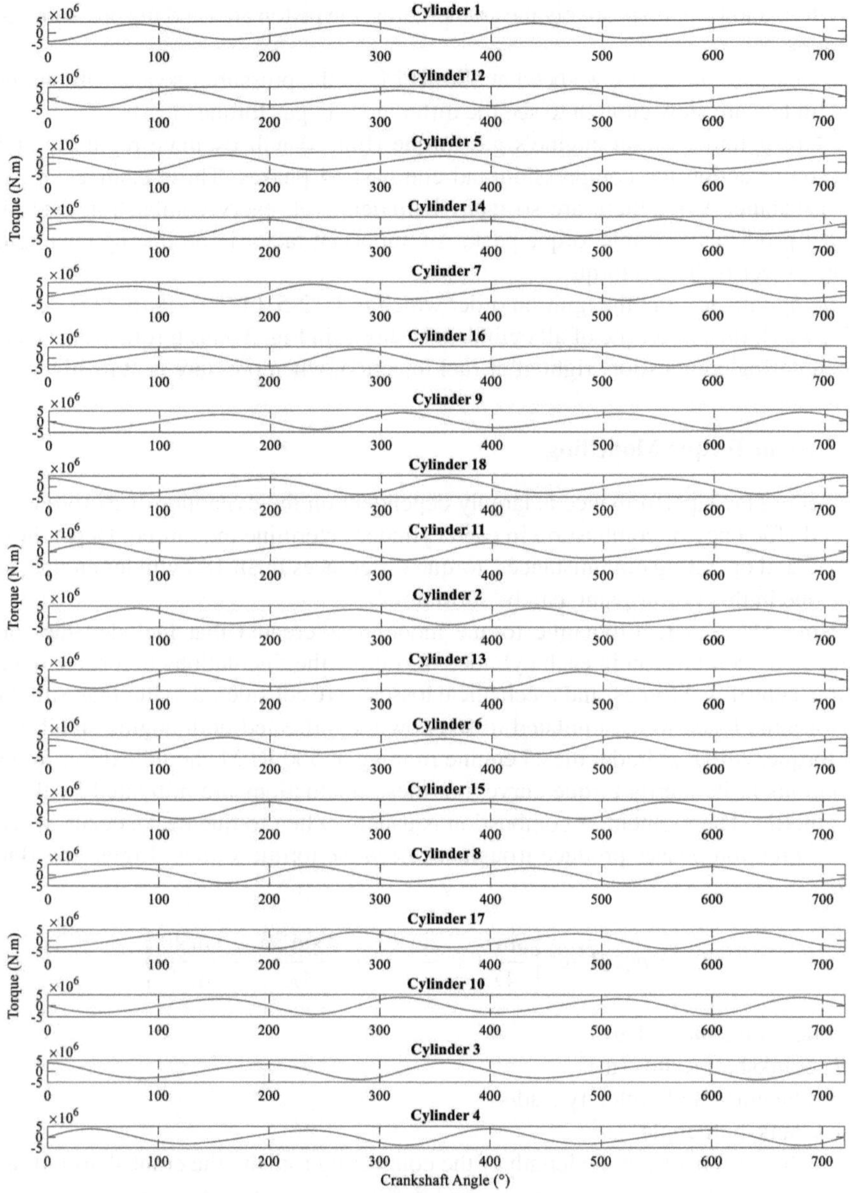

Fig. 4. Variation of inertial torque for N = 500 rpm in one engine cycle

3.4 ICE Torque Modelling

The total gas torque is the sum of gas torque of all cylinders.

$$T_g = \sum_{j=1}^{18} T_{gj} \tag{3}$$

Figure 5 shows the variation of the total gas torque in one engine cycle.

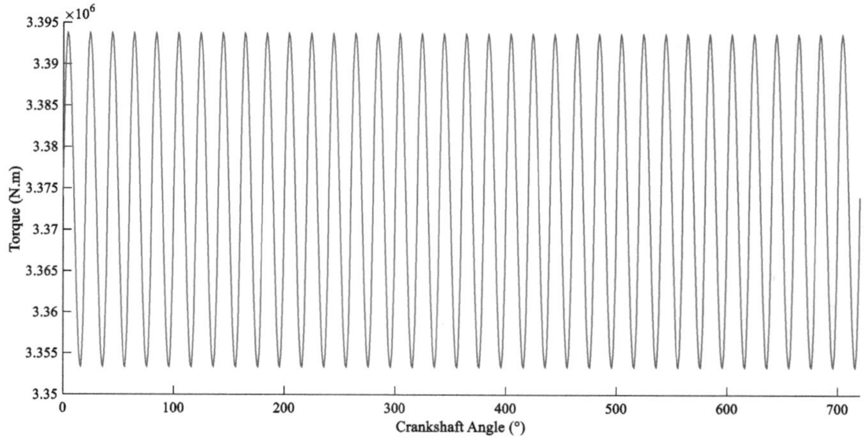

Fig. 5. Total gas torque of the Wärtsilä W18V50DF Diesel Engine

The total inertia torque is the sum of inertia torque of all cylinders.

$$T_i = \sum_{j=1}^{18} T_{ij} \tag{4}$$

Figure 6 shows the variation of the total inertia torque in one engine cycle.

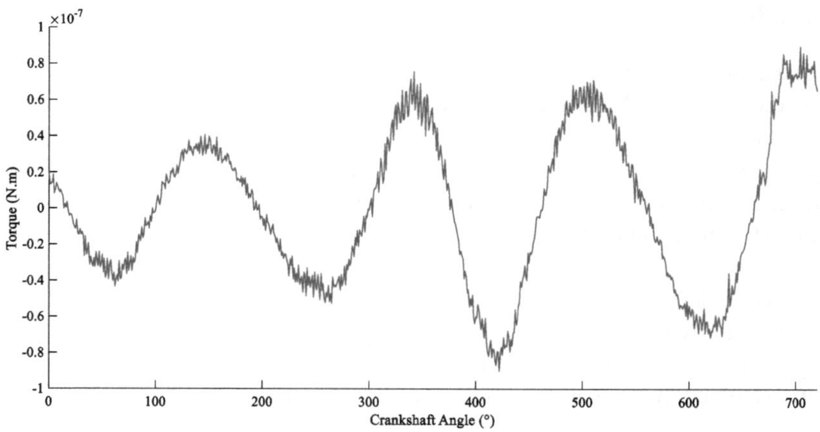

Fig. 6. Total inertia torque of the Wärtsilä W18V50DF Diesel Engine

The sum of the engine's inertia and gas torques is the engine's overall torque output. The inertia torque is a representation of the resistance and acceleration of the rotating masses, whereas the gas torque is the primary source of engine torque during combustion.

$$T_{ICE} = T_g + T_i \tag{5}$$

Figure 7 shows the total generated torque of the Wärtsilä W18V50DF Diesel Engine in one engine cycle.

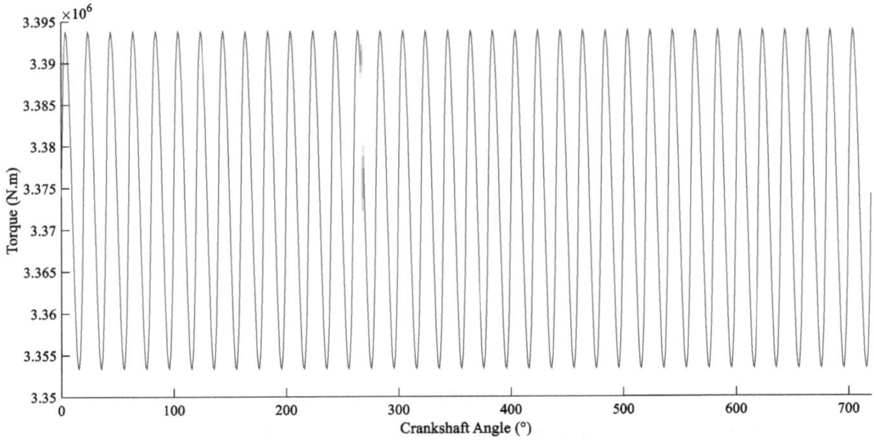

Fig. 7. Total generated torque of the Wärtsilä W18V50DF Diesel Engine

Since the Wärtsilä W18V50DF Diesel Engine is running in low regime, the inertia torque can be neglected compared to the gas torque. When the combustion gases press on the pistons during the expansion phase (Fig. 7), torque is produced. The gas torque increases with increasing gas pressure.

4 Conclusion

Using experimental pressure measurements and dynamic modeling of an 18-cylinder Wärtsilä diesel engine, Mauritania's thermal power plants may operate more efficiently and last longer.

By putting the engine's behavior into mathematical representations, dynamic modeling aids engineers in comprehending the interactions between variables like torque and pressure. This method greatly enhances the resilience and efficiency of the 18-cylinder Wärtsilä engine.

Real-world data from experimental pressure measurements validates the models and makes sure the dynamics of the engine are appropriately reflected in them. In-depth investigation of the dynamics of torque and pressure is also made possible by MATLAB simulations, which provide performance estimates for the engine.

Future forecasts and optimizations will be more reliable due to the strong correlation shown between simulation findings and experimental data. By guaranteeing increased engine reliability and enabling notable decreases in fuel consumption and pollutant emissions, this modeling technique maximizes the power plants' overall efficiency.

Acknowledgements. This work is financed by the Tunisian-Mauritanian project n° "23MT1003", supported by the Tunisian Ministry of Higher Education and Scientific Research and the Mauritanian Ministry of Higher Education through ANRSI.

References

Heywood, J.B.: Internal Combustion Engine Fundamentals. McGraw-Hill (1988)

Merker, G.P., Schwarz, C., Stiesch, G., Otto, F.: Simulating Combustion: Simulation of Combustion and Pollutant Formation for Engine-Development. Springer-Verlag (2005)

Ligier J.L, Baron E.: Acyclisme et vibrations, Applications aux moteurs thermiques et aux transmissions, Publications de l'Institut Francais de Petrole et de Moteurs, Editions TECHNIP, Paris, Vol. 1, p. 102 (2002)

Hmida, A., Hammami, A., Chaari, F., Ben Amar, M., Haddar, M.: Effects of misfire on the dynamic behavior of gasoline Engine Crankshafts. Eng. Fail. Anal. **121**, 105149 (2021)

Hmida, A., Hammami, A., Khabou, M.T., Chaari, F., Haddar, M.: Dynamic Behavior of Spur Gearbox with an Elastic Coupling Under Acyclism Regime. Advances in Condition Monitoring of Machinery in Non-Stationary Operations. Applied Condition Monitoring 2018; vol 9. Springer, Cham (2018). https://doi.org/10.1007/978-3-319-61927-9_30

Hammami, A., Hmida, A., Chaari, F., Khabou, M.T., Haddar, M.: The effect of a cracked tooth on the dynamic response of a simple gearbox with a flexible coupling under acyclism operation. J. Theor. Appl. Mech. **57**(3), 591–603 (2019)

Hammami, A., Hmida, A., Khabou, M.T., Chaari, F., Haddar, M.: Applications of ceemdan in dynamic behavior of defected spur gearbox running under acyclism regime. Journal of Mechanics (2020). https://doi.org/10.1017/jmech.2020.19

Ligier, J.-L., Baron, É.: Acyclisme et vibrations, Editions TECHNIP (2002)

Determination of the Optimum Design of One-Dimensional Wind Turbine Under the Weather Conditions of Mosul City

Yaser S. H. Mahmood[1,2] and Mohamed Taktak[1](✉)

[1] Laboratory of Mechanics, Modeling and Productivity (LA2MP), National School of Engineers of Sfax, University of Sfax, BP N °, 1173 - 3038 Sfax, Tunisia
yaseralmola@uomosul.edu.iq, mohamed.taktak@fss.usf.tn
[2] Department of Mechanical Engineering, Mosul University, Mosul, Iraq

Abstract. In this study, the response of a 1-D horizontal axis wind turbine under the quantitative and real wind velocities of the city of Mosul at three different heights, under the axes of operation and design, has been analyzed. At generator height, hourly wind velocities over a period of five years ranging between 2018 and 2023 have been subjected to statistical analysis - mean, variance, wind velocity, distribution parameters (shape and scale), and especially exceeded wind velocities - utilizing two widely used distributions. Due to the high agreement of distributions with the real data have been extracted. At these velocities, and especially their exceeded velocities, corresponding to different heights, corresponding wind velocity pressures have been calculated. Thereinafter, the behavior of the tower across the two extreme design axes has been analyzed in the load-capacity concept.

The results of the study have proved that, using the randomly generated or commonly used values in the design calculations of the minimum velocity, the probability of the wind velocity values that are, in fact, exceeded the chosen return period is high, and vice versa. From the statistical properties of the tower elements, the results have depicted that the turbine tower system design can be considered viable to two wind velocities; one from the nation's point of view, along with another from the individual customer's perspective. Depending on this, the design limits considering the concept of the relationship of the standard servicing times and the exceedance look-up tables have been recommended. With the comparison of this relationship, the agreement of the average design statistics with the national design recommendations becomes evident; where the remaining shape quality of the turbine tower structure can be considered high.

Keywords: Wind turbine · one-dimensional design · Mosul city · wind energy · optimum design · CFD simulation

A. Hammami et al. (Eds.): MOSCOSSEE 2024, ACM 23, pp. 64–79, 2025.
https://doi.org/10.1007/978-3-032-00930-2_6

1 Introduction

1.1 What is energy?

Energy is defined as the ability of a system to produce external work. Energy is the basic building block that forms the basis of every person's daily activities and is important for a sustainable quality of life. Energy is critical for social and economic growth and development. In this sense, energy is divided into mechanical energy (i.e. potential or kinetic energy) and primary and secondary energy sources according to their conversion into energy. Primary energies are hard coal, crude oil, natural gas, hydraulic energy, geothermal energy, solar energy, and are utilized through conversion. According to their composition, energy sources are divided into conventional (non-renewable) and renewable energy.

1.2 Global Energy Problems

Today, the demand for energy is increasing due to factors such as industrialization, technological development, and population growth. For these reasons, energy is seen as the most important issue in the world. Energy is no longer a local issue; it has become a global issue and has begun to guide international politics. The fossil energy complex, which we call conventional energy sources, accounts for about 90% of the energy used today. Fossil-based energies such as coal, oil, natural gas and uranium (nuclear energy) are not as clean and inexhaustible as renewable energy sources. It is known that these resources will inevitably run out one day. While there are many technological applications that run on fossil fuels today, these energy sources are expected to run out one day, and countries will be greatly affected by this crisis, primarily economically. In addition to the economic effects of the global energy crisis, it also causes environmental issues due to the environmental damage it causes. Most of the environmental pollution we see today is caused by fossil fuels. The harmful gases emitted by the use of fossil fuels are not only harmful to human health, but also cause environmental pollution. This environmental pollution arises against the backdrop of major environmental issues such as rising water temperatures, which we call global warming, acid rain, depletion of the ozone layer, and the greenhouse effect.

1.3 Renewable Energy Sources

When events such as global warming, the global oil crisis, as well as geographical conditions, increased environmental awareness, and limited use of fossil fuels occur, the importance of renewable energy sources grows and the trend towards this sector is cited as essential. Renewable resources are characterized as solar energy, geothermal energy, hydraulic energy, wind energy geothermal energy, biomass, waves, and non-fossil energy sources.

Renewable energy is also described as energy production that can be produced from natural sources and continuously regenerates itself, as it can automatically self-renew itself and does not perish, unlike other forms of energy. In addition, since renewable energy sources are local resources that are environmentally favorable (reducing carbon emissions), there is no need to import them. For this reason, it is particularly important in terms of reducing foreign energy dependence and protecting the environment (Fig. 1).

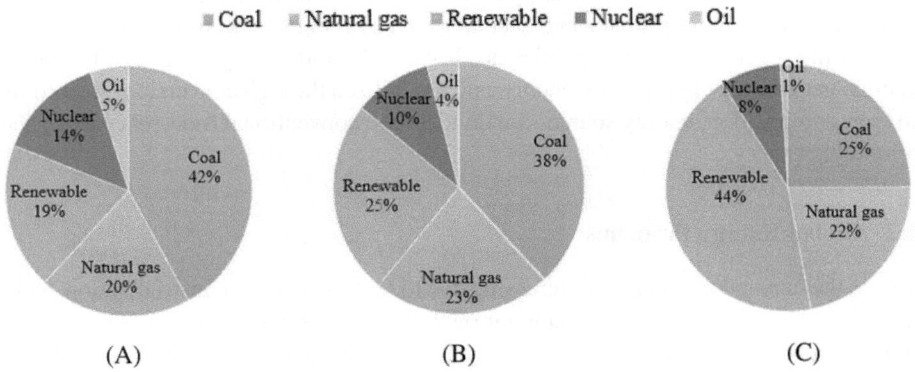

Fig. 1. Global primary energy sources for electricity generation (a) for 2010 (b) for 2017 (c) announced policies for 2040 (Energy Agency, 2021)

Renewable energy encompasses a variety of sources that rely on being naturally renewable without unsustainably depleting natural resources. The main types of renewable energy include:

1. Solar energy: Solar radiation is used to generate electricity and heat water. This is done by using solar cells that convert sunlight directly into electricity.
2. Wind energy: Electricity is generated using wind through wind turbines. Wind energy is one of the fastest growing types of renewable energy.
3. Hydroelectric power: Rivers and dams are used to generate electricity through the movement of water. Dams and hydroelectric plants are among the oldest and most widely used forms of renewable energy.
4. Geothermal energy: Uses the earth's internal heat to generate electricity. This is done by using a thermal well in which hot water vapor or hot water underground is extracted.
5. Marine energy: Uses the movement of waves and ocean currents to generate electricity. This technology includes wave elevators, wave engines, and marine current engines.
6. Plant energy: This category includes the use of organic waste and biodegradable plant materials to generate biogas that can be used for power generation.

It plays a big role in the future as it is sustainable because it regenerates from natural processes. After the industrial revolution in the 1970s and the widespread use of fossil energy resources that negatively impacted the environment leading to the emergence of some undesirable phenomena such as desertification, drought, melting ice and other signs of climate change. Therefore, it was necessary to turn to alternative sources of renewable energy, which are characterized by having little negative impact on the environment, the simplest of which is wind energy [1].

1.4 Governorate Location and Wind Direction

Nineveh Governorate is located in the northwestern part of Iraq. It occupies a large area of (37,597.53) square kilometers of the total area of Iraq, or 8.64% of the area of Iraq. It is bordered to the north by Dohuk Governorate, to the west by the Syrian Arab Republic, to the east by Erbil Governorate, and to the west by Salahuddin Governorate and Anbar Governorate The study area extends from latitude 340:55’ north to latitude 370:03’ north, and from longitude 410:25’ east to longitude 440:25’ east.

The governorate’s climate varies according to the surface topography. Nineveh Governorate is characterized by the diversity of its terrain, which includes plateaus, hills, and undulating plains, some of which represent the southern boundary of the undulating region. It can be seen from Fig. 2 below that the elevation above mean sea level in the governorate ranges from slightly steep to highly steep (69–1646 m), where the directions of the highest elevation are generally towards the northwest/southeast, except for Mount Sinjar, where the direction is from east to west.

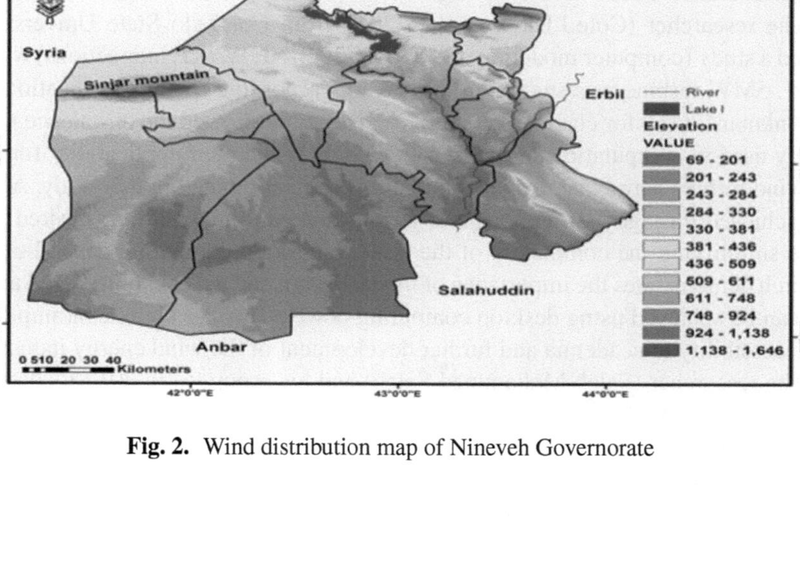

Fig. 2. Wind distribution map of Nineveh Governorate

In this present study the response of a 1-D horizontal axis wind turbine under the quantitative and real wind velocities of the city of Mosul at three different heights, under the axes of operation and design, has been analyzed.

1.5 Literature Review

The researchers (Alpinar) and his colleague in (2006) [2] studied the evaluation of wind turbine characteristics and wind energy characteristics for electricity production This study is an analysis of wind turbine characteristics and wind energy characteristics for four regions in Turkey, namely Maden, Agin, Elazig and Keban. They analyzed the characteristics of wind turbines in these regions through wind speed and wind direction data, analyzed the six years from 1998 to 2003, and extracted estimates of wind energy production and power generation for the six turbines, the study showed that the power of the turbines ranges between (300 kW) and (2300 kW), and they concluded through their research that the appropriate site for installing wind turbines is the site with the highest wind frequency.

In 2008, the researcher (CAMELIA) and his group [3] presented a theoretical study, which is modeling, simulation and control of wind turbines in this study The turbine system consists of four elements: Wind, wind turbine, alternator and controller. They are the important elements of a wind turbine, which must be analyzed, modeled and simulated, and the main objective of this study was to define the mathematical model of a wind turbine, to validate it through simulation, and to design a suitable controller to provide a general viewpoint regarding the domestic use of this type of clean energy production. Every part of the wind turbine has been analyzed, modeled and simulated. And in the end, these elements are linked together and the whole system is modeled and simulated. The simulation results confirm the validity of the developed mathematical models and can be used to better design these systems.

The researcher (Cole.J.Daves) [4] in 2012 from Colorado State University conducted a study (computer modeling of WIND TURBINE WAKE interactions) where the NREL 5MW turbine was studied in SolidWorks according to the specifications given in (Jonkman. 2009) for classified conditions. The study showed the turbulence methods widely used in computational fluid dynamics (CFD) and their applicability for modeling wind turbine aerodynamics, all simulations were computed in this study. Accuracy was achieved through assumptions that were designed to maintain the desired physics while simplifying the complexity of the issue to desktop computing capabilities. This research demonstrates the importance of model design and the capabilities and accuracy that can be achieved using desktop computing power. This has significant implications for accessibility to academia and further development of the wind energy industry.

The researcher (Saleh Mohammed Saleh) and his group [5] in 2012 focused in his study on analyzing the performance of wind turbine systems under the influence of different factors, this research used simulation models to study small power systems based on different weather conditions. The results were drawn using Matlab to analyze the performance of two wind turbines: Whisper-500 3.2 KW and NY-WSR1204 600W, and this study provided a useful simulation model for practical use. According to MATLAB among the simulation program, there is a significant influence of air characteristics on the mechanical, electrical and power generation of a wind turbine. These include factors such as turbine blade, wind speed, air density, pressure, temperature, and power factor. The study shows that environmental factors have a tremendous impact on power generation, where wind speed and density lead to increased output and vice versa for temperature. Which will lead developers and researchers to focus on them with the highest priority.

In 2012, Zijun Zhang [6] conducted a study to improve the performance of wind turbines through effective control strategies to reduce the cost of power generation is highly desired by the wind industry. The majority of the literature on wind turbine performance has focused on models derived from fundamental principles of physics. Physics-based models are usually complex and inaccurate due to the fact that wind turbines include mechanical, electrical, and software components. These components interact with each other and are subjected to variable loads generated by the wind as well as the rotating elements of the wind turbine. Recent advances in data acquisition systems allow for the collection of large amounts of wind energy data. Although the primary purpose of data collection is to observe wind turbine conditions, the collected data provides a golden opportunity to address the most challenging issues for wind turbine systems. In this study data mining is applied to build accurate models based on the collected data on turbines. To solve the data-driven models, evolutionary computation algorithms were applied. The optimization of wind turbines with different objectives was studied to achieve different research goals and two research directions of wind turbine performance were pursued, optimizing wind turbine performance and optimizing wind farm performance. To optimize the wind farm performance is to minimize the total cost of wind farm operation based on the calculated turbine scheduling strategies. The methodology presented in the thesis is applicable to operations besides the wind industry.

Rodrigues and Rossi [7] in 2016 compared electricity production from small wind turbines using HOMER, where wind turbines were simulated through HOMER from the perspective of wind resources in three different cities, two in Brazil and one in the U.S. The results indicate that energy costs depend on the characteristics of wind turbines (technical and operational characteristics of wind turbines), so that maximum energy production can be achieved.

In 2016, Wiratama and his group [8] conducted a study to investigate the factors that affect wind turbine capacity and the study aims to investigate the factors that affect the enhancement of energy capture capabilities of wind turbines in Indonesia, these factors include wind speed, air density, the study shows that these mentioned factors will help developers and researchers focus on the selection or manufacture of wind turbines and the researchers found that air density is directly proportional to the power output of the wind turbine by holding other factors constant as well as the speed is proportional to the cube of the speed. These factors can change with altitude and temperature. In addition, the power and power output curves increase as the site temperature decreases.

While the researcher (Samah Shi'ah and Mohammed Abdullah [9] in 2016 conducted a preliminary survey of wind energy in Tuz Khurmatu, the aim of this work was to study the possibility of building a wind farm in (Tuz Khurmatu) in Iraq and at different altitudes, and the best option to create the proposed wind farm from turbines (Nordex N29), (Nordex N43) and (Enercon-33), the study showed that wind speeds range from m/s(3) and (m/s3.14). (14) and (m/s3.26) at an altitude of (m)10, and at an altitude of (m40) the speed was (4.86 m/s). The study site in Tuz-Khormato is a low wind speed site suitable for low-speed wind turbines.

In 2017, the researchers (Roshin Ahmed and Mahdi Abdul Hussain) [10] studied the modeling and simulation of wind turbines using Matlab-Simulink. In this research, a mathematical model was developed and the factors that affect the production of electrical energy generated by wind turbines were studied. These factors are wind speed which is influenced by temperature which causes air density change and leads to wind speed variation, power factor as a function of pitch angle and blade tip speed where wind turbines are directly related to wind speed and tip speed and modeling and simulation technique plays a big role in the design and analysis of these wind turbines. In the past years, the demand for Matlab-Simulink has been one of the most common software applications, which is important for modeling and simulation of dynamic systems. Because it provides a graphical interface, easy to access, design, build and verify mathematical models as well as from the results that can be obtained in this study Wind speed variation with time, wind turbines are directly related to wind speed and tip speed, air density has a great influence on wind turbine performance. The power available in the wind is directly proportional to the air density also as the air density increases.

n 2018, Fawaz Saleh Abdullah and his group [11] studied the modeling and simulation of a 1.5 MW wind turbine, which is a variable-speed turbine, which is the dominant type in recent years. There are usually two control strategies for variable-speed wind turbines. In the case of low wind speed below the rated value, the speed controller can continuously adjust the rotor speed to maintain the speed at the level, giving the maximum power factor, and then the efficiency of the turbine is maximized. The vane angle needs to be regulated in conditions above the rated wind speed when the rotational speed remains constant. Small changes in vane angle can affect the output power. This study showed the relationship between the power output of the turbine and the wind speed as well as the turbine speed and the results showed that the power output of the wind turbine is according to the wind turbulence. Matlab/Simulink is used to simulate this turbine. This study identified one type of wind turbine with a rated power of 1.5 MW.

The researcher (Esraa Adel Mansour) [12], in 2018, conducted a practical study on evaluating the performance of a horizontal axis wind turbine under the influence of real site conditions, where this thesis showed a theoretical and applied analysis of the possibility of using wind turbines in Kirkuk city as a clean and environmentally friendly means of energy, and to study the effect of weather conditions on wind turbine performance, namely temperature, wind speed, wind direction, and dust density. The conversion of wind kinetic energy into mechanical energy was studied by a horizontal wind turbine in Kirkuk city at a height (16 m) from the ground with three blades with a capacity (400 W) and connected to an electrical system made to measure the voltage and current of the wind turbine, taking readings for summer and winter from meteorological information, and analyzing the data using Matlab software to find the theoretical and measured capacity with practical results and find the efficiency of the system and its applicability to generate electricity for domestic use. The results showed that the summer season is better than the winter season, where the monthly average power output during the summer amounted to hr. 100724 kw, while during the winter, the monthly energy rate reached (9552.5 kw.hr), due to the low winter wind speed. The wind speed during the summer season (w/m 4) and during the winter season (w/m 1.87), where the study showed through the results that the effect of high temperatures and humidity is inversely proportional to the power generated by the turbine. As for dust, it has relatively little effect on the performance of the wind turbine. The power coefficient reached (0.37) by comparing the practical and theoretical results. A polynomial relationship was found between Wind speed and the power generated by the wind turbine.

Most of the studies investigate the wind maps, while some of the literature addresses the effect of air humidity and wind fluctuation on the efficiency of the turbine. Some of the very recent research using the measured wind parameters was on the improvement in technology in the production of clean electricity from wind power. A gap in all of this literature is in the allocation of the optimal design of the turbines given the local site conditions. Of the research found, it appears, and to the best knowledge of the authors, that the production of wind turbines was an import from outside the Arab world from countries that are leaders in the technology of wind turbines. Most of the wind rotor models are of different designs and, in general, appear to be two-dimensional models. Sudden advancements in computation technology and the evolution of explaining wind energy conversion theories saw the advent of one-dimensional models as well as the design power coefficient being considered for the design of one-dimensional turbines. As such, the physical mode of turbine constructs has had links with the technology and methodology of the time. The design model tackled in this study is one-dimensional.

2 Methodology

Also, an optimization technique is used in order to obtain a design of the one-dimensional wind turbine according to the weather conditions of the city. The mathematical equation for the design of the one-dimensional wind turbine is optimized by minimizing the cost of the energy produced by the wind turbine.

It is found that the use of techniques constitutes a unique methodology for the start, the conjunction of the model with the real design, and the final comparison between theoretical and practical study. The methodologies used enhance the systematic approach of the research to verify the design of the wind turbine. The performance characteristics, the output power, and the coefficient of power of the wind turbine are obtained by applying an analytical and computational model. This section outlines the methodology used in the design of the wind turbine. Firstly, a detailed data collection process is presented, and then the fundamental framework for the numerical study is provided with some key factors that should be considered. This is followed by details of the optimization technique and the appropriate analytical and computational models used in this research.

2.1 Theoretical Analysis

The area of the rotation surface expresses the area arising from the rotation of the blades, which is determined by the length of the feathers, the greater the lengths of the blades, the greater the area of the rotation surface, and thus the volume of air that hits the plane of rotation, and the area of the rotation surface is also called the diameter of the turbine and the area of the rotation surface in the turbine and is calculated from the following equation:

$$A = \pi\frac{D^2}{4} \tag{1}$$

where D represents the diameter of the rotor in the turbine. The diameter of the rotor and the amount of energy that can be obtained from the turbine is proportional to the area of the rotor, which in turn is proportional to the square of the diameter [13].

The mathematical equations used to study the performance of conventional wind turbines are as follows:

$$\text{Kinetic energy(K.E)} = \frac{1}{2}mu^2 \tag{2}$$

$$\text{power wind turbin(W)} = P_w = \frac{K.E}{t} = \frac{\frac{1}{2}mu^2}{t} \tag{3}$$

$$\text{Density } \rho\left(\frac{Kg}{m3}\right) = \frac{m}{V} \tag{4}$$

$$\text{Volume}\left(m^3\right) = \text{Area} * \text{Length} = V = A * L \tag{5}$$

MASS = Density* Volume

$$M(Kg) = V * \rho * A * L \tag{6}$$

$$\text{Velocity} = \frac{\text{Length}}{\text{time}} = u = \frac{L}{t} \tag{7}$$

Substituting Eq. (6) and Eq. (5) into Eq. (2) to get the following equation which represents the wind power:

$$P_W = \frac{1}{2}\rho A\frac{L}{t}u^2 = \frac{1}{2}\rho Au^3 \tag{8}$$

By the 1st law of thermodynamics we get this equation:

$$PV = mRT \tag{9}$$

We can calculate the density of air from this eq.:

$$\rho = \frac{P}{RT} \tag{10}$$

where: R = 287J/kg.s

Reynold's number Re_ is represented by the following equation:

$$R_e = \frac{Uc\rho}{\mu} \tag{11}$$

U is the wind velocity
where (CP) is the power factor

$$C_P = \frac{P_R(\text{Rotor Power})}{P_w(\text{Wind Power})} \tag{12}$$

where (CL) is the lifting coefficient and is calculated from the following relationship:

$$C_L = \frac{2F_L}{\rho U^3 A} \tag{13}$$

where (FL) is the lifting force

The braking coefficient (CD) is calculated from the following relationship:

$$C_D = \frac{2F_D}{\rho U^3 A} \tag{14}$$

Tip - Speed Ratio: It is the ratio between the speed at the end of the vane "Tip Blade" and the wind speed, which increases as the length of the vane increases, is an important factor in wind turbine design. TSR will help in knowing the turbine tip speed ratio to maximize the power output and efficiency of the wind turbine. If the turbine blades rotate too slowly, too much wind will pass through the gaps between the blades instead of giving power to the turbine. If the blades spin too fast, they can create too much turbulent air or act as a solid wall against the wind. Therefore, an optimal angular speed exists at which maximum energy extraction is achieved [15].

The vane tip speed was obtained by the following equation:

$$\lambda = \frac{\omega R}{U} \tag{15}$$

The power factor based on the velocity at the tip of the vane can be represented by the following relationship:

$$K_P = \frac{Cp}{\lambda} \tag{16}$$

2.2 Data Collection

In order to conduct comprehensive research in the field of wind speed evaluation of the location Mosul, several climatic data should be studied. This information includes the annual wind speed, the monthly wind speed, and the thermal ranges of the measured period. These data were obtained either from the local meteorological stations or were determined from the historical data recorded at a certain height of each month. These relations should be expressed mathematically as a mathematical model, in order to help either in obtaining the weather data of the mean month from the monthly data or obtaining the monthly data from the values of the mean month data. The location data has been obtained, which shows latitude and where Iraq is approximately located.

This comprehensive study will be conducted to determine the design and dimensions of the optimum design for one-dimensional turbines in the weather region of Mosul, the capital of Nineveh governorate. With the availability of detailed climate data in this region, particularly wind speed at different heights, temperature, and relative humidity of the air, the required initial data for the model is now available, and it has been verified to be a good response. The time of the sun slows down with the location of the months will be presented to crystallize the development of a timetable for the slowdown of the period of the sun, through which we will focus on manufacturing the models, the flow path required by these models, and data collection. The speed of the wind has been collected from a friend, a graduate in the field of energy engineering from the College of Engineering at Nineveh University.

With respect to each of the sites, the speed of the wind, the income, the amount of moisture held by the air, and the heat periods have been addressed, and their properties of the month have been addressed on functioning completely. Verification of the data was carried out by examining data continuity, accuracy, and by resolving any data inconsistency. The collected data was also analyzed and modeled to support the research. The primary information required in order to develop the model has been obtained. That information included: (i) the time of sunrise, sunset, and the length of the daylight time, and (ii) the wind speed, and (iii) the temperature of the air (Table 1).

Table 1. XXX

Parameter	Value
Rotor radius(R)	1
The divided parts of the Blade (r_i)	20
Design wind Speed	8
Design tip speed ratio(λ)	6
Rotational speed (ω')	48
Number of Blade (N_b)	3
Rotor speed (Ω)	458
Reynolds number (Re)	1280000
Design angle of attack (α)	6.75^O
Design lift coefficient (Cl)	1.109
Design drag coefficient (Cd)	0.00966
Airfoil File Name	S827

R	ri	Ne	Nb	λ	Vi	α	Cl	$\lambda_{r,i}$	ϕi	C_i	ϕi (deg)	theta θ	W red/s	Rpm	Vt	V_{rel}	R_{ei}	RE
1	0.05	20	3	6	8	6.75	1.109	0.3	0.85289	0.7751	48.86717	42.1172	48	458.366	37.2049	38.0553	3.6871	1.95E+06
1	0.1	20	3	6	8	6.75	1.109	0.6	0.68692	1.02743	39.3575	32.6075	48	458.366	49.3167	49.9614	6.41649	3.40E+06
1	0.15	20	3	6	8	6.75	1.109	0.9	0.55865	1.0331	32.00853	25.2585	48	458.366	49.5886	50.2298	6.48653	3.43E+06
1	0.2	20	3	6	8	6.75	1.109	1.2	0.46316	0.95455	26.53705	19.787	48	458.366	45.8186	46.5118	5.54975	2.94E+06
1	0.25	20	3	6	8	6.75	1.109	1.5	0.392	0.85908	22.46005	15.71	48	458.366	41.236	42.0049	4.51071	2.39E+06
1	0.3	20	3	6	8	6.75	1.109	1.8	0.33807	0.76926	19.36974	12.6197	48	458.366	36.9243	37.781	3.63291	1.92E+06
1	0.35	20	3	6	8	6.75	1.109	2.1	0.29628	0.69084	16.97556	10.2256	48	458.366	33.1605	34.1119	2.94575	1.56E+06
1	0.4	20	3	6	8	6.75	1.109	2.4	0.26319	0.62401	15.07991	8.32991	48	458.366	29.9525	31.0025	2.41823	1.28E+06
1	0.45	20	3	6	8	6.75	1.109	2.7	0.23647	0.56732	13.54876	6.79876	48	458.366	27.2315	28.3823	2.01273	1.07E+06
1	0.5	20	3	6	8	6.75	1.109	3	0.2145	0.5191	12.28997	5.53997	48	458.366	24.9166	26.1694	1.69806	8.99E+05
1	0.55	20	3	6	8	6.75	1.109	3.3	0.19616	0.47782	11.23893	4.48893	48	458.366	22.9353	24.2905	1.4508	7.68E+05
1	0.6	20	3	6	8	6.75	1.109	3.6	0.18063	0.44223	10.34941	3.59941	48	458.366	21.2268	22.6843	1.25395	6.64E+05
1	0.65	20	3	6	8	6.75	1.109	3.9	0.16734	0.4113	9.587596	2.8376	48	458.366	19.7426	21.3018	1.09519	5.80E+05
1	0.7	20	3	6	8	6.75	1.109	4.2	0.15583	0.38424	8.928332	2.17833	48	458.366	18.4435	20.1038	0.96559	5.11E+05
1	0.75	20	3	6	8	6.75	1.109	4.5	0.14578	0.36039	8.352538	1.60254	48	458.366	17.2987	19.059	0.85858	4.55E+05
1	0.8	20	3	6	8	6.75	1.109	4.8	0.13693	0.33923	7.845526	1.09553	48	458.366	16.2832	18.1423	0.76931	4.07E+05
1	0.85	20	3	6	8	6.75	1.109	5.1	0.12908	0.32035	7.395815	0.64582	48	458.366	15.377	17.3336	0.69411	3.67E+05
1	0.9	20	3	6	8	6.75	1.109	5.4	0.12207	0.30341	6.994318	0.24432	48	458.366	14.5639	16.6165	0.63021	3.34E+05
1	0.95	20	3	6	8	6.75	1.109	5.7	0.11578	0.28814	6.633751	-0.11625	48	458.366	13.8306	15.9777	0.57547	3.05E+05
1	1	20	3	6	8	6.75	1.109	6	0.1101	0.27429	6.308215	-0.44179	48	458.366	13.1661	15.4061	0.52823	2.80E+05

span1			span2			span3			span4			span5			span6		
x	y	z	x	y	z	x	y	z	x	y	z	x	y	z	x	y	z
0.7751	0	0	1.02743	0	0	1.0331	0	0	0.95455	0	0	0.85908	0	0	0.76926	0	0
0.7719	0.00053	0	1.02318	0.00071	0	1.02882	0.00071	0	0.95061	0.00066	0	0.85553	0.00059	0	0.76608	0.00053	0
0.76269	0.00268	0	1.01098	0.00356	0	1.01656	0.00358	0	0.93927	0.0033	0	0.84533	0.00297	0	0.75694	0.00266	0
0.74846	0.00689	0	0.99211	0.00913	0	0.99758	0.00918	0	0.92174	0.00848	0	0.82955	0.00763	0	0.74281	0.00683	0
0.73021	0.01299	0	0.96792	0.01722	0	0.97326	0.01732	0	0.89926	0.016	0	0.80932	0.0144	0	0.7247	0.01289	0
0.70868	0.02049	0	0.93938	0.02716	0	0.94456	0.02731	0	0.87275	0.02524	0	0.78546	0.02271	0	0.70333	0.02034	0
0.68424	0.02883	0	0.90699	0.03822	0	0.91199	0.03843	0	0.84266	0.03551	0	0.75838	0.03196	0	0.67908	0.02861	0
0.65708	0.03779	0	0.87099	0.05009	0	0.8758	0.05036	0	0.80921	0.04653	0	0.72828	0.04188	0	0.65213	0.0375	0
0.62774	0.04722	0	0.8321	0.06259	0	0.83669	0.06294	0	0.77308	0.05815	0	0.69576	0.05234	0	0.62301	0.04687	0
0.59677	0.05678	0	0.79104	0.07527	0	0.7954	0.07568	0	0.73493	0.06993	0	0.66143	0.06293	0	0.59227	0.05635	0
0.56464	0.06604	0	0.74845	0.08754	0	0.75258	0.08802	0	0.69536	0.08133	0	0.62581	0.07319	0	0.56038	0.06554	0
0.53177	0.07451	0	0.70488	0.09877	0	0.70877	0.09932	0	0.65488	0.09177	0	0.58939	0.08259	0	0.52776	0.07395	0
0.49852	0.08166	0	0.66082	0.10824	0	0.66446	0.10884	0	0.61394	0.10056	0	0.55254	0.0905	0	0.49476	0.08104	0
0.46496	0.08655	0	0.61633	0.11473	0	0.61973	0.11536	0	0.57261	0.10659	0	0.51534	0.09593	0	0.46146	0.0859	0
0.43015	0.08924	0	0.57018	0.11829	0	0.57332	0.11894	0	0.52973	0.1099	0	0.47675	0.09891	0	0.4269	0.08856	0
0.39436	0.09035	0	0.52274	0.11976	0	0.52563	0.12042	0	0.48566	0.11126	0	0.43709	0.10013	0	0.39139	0.08966	0
0.35802	0.09008	0	0.47458	0.1194	0	0.47719	0.12006	0	0.44091	0.11093	0	0.39682	0.09984	0	0.35532	0.0894	0
0.32155	0.08857	0	0.42623	0.1174	0	0.42858	0.11804	0	0.396	0.10907	0	0.35639	0.09816	0	0.31913	0.0879	0
0.28534	0.08591	0	0.37823	0.11388	0	0.38031	0.11451	0	0.3514	0.1058	0	0.31625	0.09522	0	0.28319	0.08526	0
0.2498	0.08222	0	0.33112	0.10899	0	0.33294	0.10959	0	0.30763	0.10126	0	0.27686	0.09113	0	0.24791	0.0816	0
0.21531	0.0776	0	0.2854	0.10286	0	0.28697	0.10343	0	0.26515	0.09556	0	0.23863	0.086	0	0.21368	0.07701	0
0.18227	0.07215	0	0.2416	0.09563	0	0.24293	0.09616	0	0.22446	0.08885	0	0.20201	0.07996	0	0.18089	0.0716	0
0.15103	0.06598	0	0.2002	0.08745	0	0.2013	0.08794	0	0.186	0.08125	0	0.1674	0.07312	0	0.14989	0.06548	0
0.12196	0.0592	0	0.16167	0.07847	0	0.16256	0.07891	0	0.1502	0.07291	0	0.13518	0.06562	0	0.12104	0.05876	0
0.09537	0.05194	0	0.12642	0.06885	0	0.12712	0.06923	0	0.11746	0.06396	0	0.10571	0.05757	0	0.09466	0.05155	0

The twist angle θ is the difference between the airflow angle and the design angle of attack in each blade element. The distribution of the chord and the optimal wave angle for each blade element is shown in Figs. 3 and 4.

2.3 Optimization Techniques

Studies integrated into the enhancement of wind turbine design to improve efficiency and maximize power have been developed using various optimization techniques. Complex design decisions such as the vane, generator's electrical power, dimensions, and aerodynamic shape of rotor blades need to be taken into account. Moreover, the best input/output values must be derived for a specified system to achieve the best performance of the turbine at a given wind speed. These are just some of the various design optimizations that must be undertaken to operate under the given climate conditions of Mosul city most efficiently, and every technique to approach this will have its pros and cons. Every design is optimized for only a specific purpose, in this case, to conform to the climate conditions of Mosul city only, not to generalize all designs. Several successful and unsuccessful parameters are tested to draw a conclusion. Depending on the conducted tests, the mathematical and computational techniques used in this study are analyzed to present strengths and weaknesses. Gradient-based methods, genetic algorithms, and particle swarm optimization can be used to conduct this optimization study. In general, these techniques aim to optimize a multi-variable design that can be evaluated

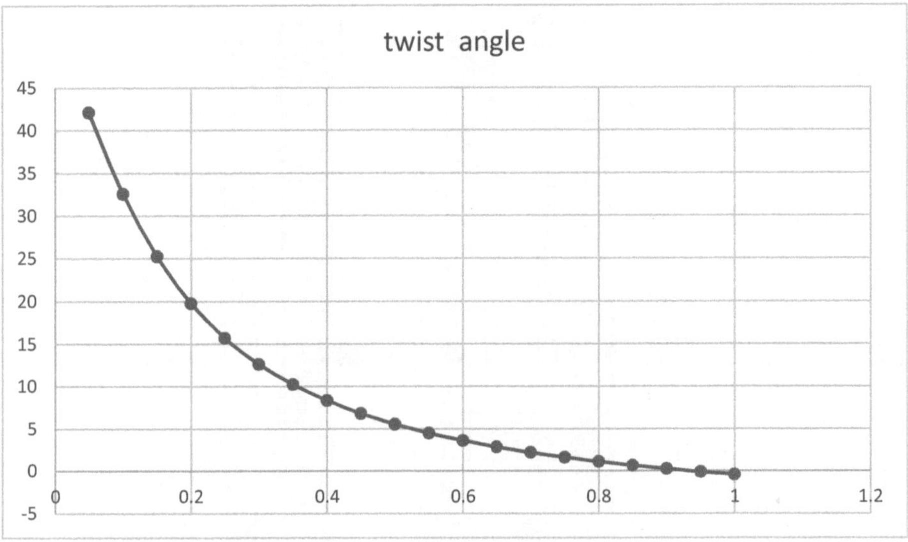

Fig. 3. Twist angle distribution plot along the length of the blade

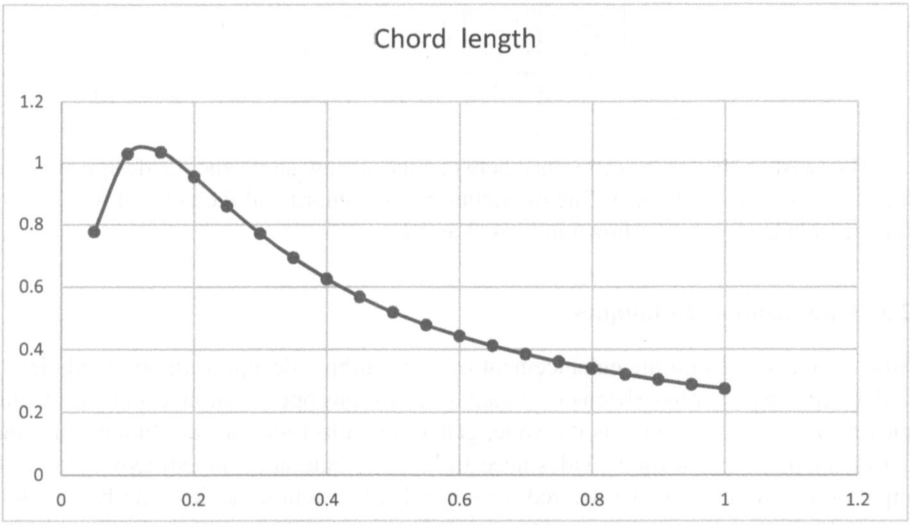

Fig. 4. Chord length distribution plot along the length of the blade

numerically and are basically variations of the same theme associated with differential evolution and random probability.

2.3.1 Optimum Blade Length

The ideal blade length considering wind characteristics of Mosul was found to be 25 m. Longer blades had the potential to capture more energy, but the cost-to-benefit ratio decreased due to mechanical stress and material costs.

2.3.2 Blade Pitch and No. of Blades

The performance was optimum for a blade pitch of 12° and a three-blade configuration. This configuration maximized efficiency at wind speeds of 3–4 m/s, which are common in Mosul.

2.3.3 CFD Simulation Results

The CFD simulations have shown that the new design achieves a very high pressure coefficient close to the practical limit, amounting to 0.44, which gives an average power output of 50 kW for the most frequent wind speeds. This constitutes a major step ahead in the performance of traditional designs, less optimized for low wind.

3 Discussion

The optimized design agreed well with the moderate wind conditions in Mosul, where achieving high efficiency at the lower wind speeds is of prime importance. It exhibited that concerning blade geometry, length, and pitch angle, captured energy increases considerably. This designed model can act as a template for wind turbines in other regions of similar climates.

The data also portray local wind turbine design. Since it is not possible for a general wind turbine design to perform under all types of weather conditions, tailoring the turbines into local environments, as the study portrays, can increase energy efficiency and cost-effectiveness.

By comparing the results we obtained above, we conclude that the length of the vane and the height of the tower have a direct impact on the performance of the wind turbine, especially the length of the vane because it will affect the value of the Reynold number because the length of the cross-section of the vane will enter the calculations of the Reynold number and thus this matter will be reflected in the lifting coefficient, which will increase with the increase in the Reynold number and the braking coefficient, which is in an opposite state. As we mentioned in the first chapter, the horizontal turbine depends on the lifting force in the rotation process, in addition to that, we notice the big difference in the produced power and its coefficient, which is also related to the value of Reynold's number.

References

1. RCREEE Regional Center for Renewable Energy and Energy Efficiency
2. Akpinar, E.K., Akpinar, S.: An Assessment of Wind Turbine Characteristics and Wind Energy Characteristics for Electricity Production. Energy Sources Part Recovery Util Environ. Eff. **28**(10), 941–953 (2006). https://doi.org/10.1080/00908310600718817

3. Leba, M., Pop, E., Barbu, C., Pop, M.: Modeling, simulation and control of wind turbine, pp. 97–102 (2008)
4. Davis, C.J.: Computational modeling of wind turbine wake interactions. Text, Colorado State University (2011). Accessed: 22 Nov. 2021. https://mountainscholar.org/handle/10217/65318
5. Salih, S.: Performance analysis of wind turbine systems under different parameters effect. Int. J. Energy Environ. IJEE **3**, 895–904 (2012)
6. Zhang, Z.: Performance optimization of wind turbines (2021)
7. Rodrigues, R.V., Rossi, L.A.: Performance of small wind turbines: simulation of electricity supply to loads connected to the public or isolated grid. Eng. Agríc. **36**, 281–290 (2016). https://doi.org/10.1590/1809-4430-Eng.Agric.v36n2p281-290/2016
8. Wiratama, K., Mara, I.M., Nuarsa, I.: Investigation of factors affecting power curve wind turbine blade. J. Eng. Appl. Sci. **11**, 2759–2762 (2016)
9. Samah, S., Oudah, S., Salah, M., Author, C., Samah, Oudah, S.: Preliminary investigation study of wind energy at Tuz-Khormatu in Iraq. *Private researcher in wind energy projects. **Ministry of Electricity/Iraq (2016)
10. Tariq, R., Hamdi, A., Abdul Hussein, M.: Modeling and Simulation of Wind Turbine Generator Using Matlab-Simulink. J. Al-Rafidain Univ. Coll. **40**, 282–300 (2017)
11. Ibrahim, M., Fawaz, S., Abdullah, Salih, B.: Modeling and Simulation of 1.5MW Wind Turbine (2018)
12. Adil Mansoor, E.: تقييم أداء توربين ريحى افقى المحور تحت تأثير ظروف الموقع الحقيقية," 2018.
13. Burton, T., Sharpe, D., Jenkins, N., Bossanyi, E.: Wind Energy Handbook (2002)
14. Sitthichok Sitthiracha, M.R.: An Analytical Model of Spark Ignition engine for performance prediction. Mongkut's Institute of Technology North Bangkok, King (2006)
15. Dhariwal, H.P., Yadav, D., Roy, B.K.: PREDICTION THE EFFECT OF TIP SPEED RATIO ON WIND TURBINE GENERATOR OUTPUT PARAMETER (2012)
16. Schubel, P., Crossley, R.: Wind Turbine Blade Design, pp. 1–34 (2014)
17. Shehab, Z.N., Farhhan, A.F., Faisal, R.M.: Spatial variation influence of landscape patterns on surface water quality across an urbanized watershed in Mosul city, Iraq. Sustainable Water Resources Management **10**(5), 181 (2024). researchgate.net
18. Saleh, R.A.Q., Oleiwi, A.S., Lateef, Z.Q.: The Impact of Climate Change on Water Quality and Consumption in the Tigris River Basin (Mosul-Baghdad). IOP Conference Series: Earth and Environmental Science, Vol. 1374. No. 1. IOP Publishing (2024). iop.org
19. Qubaa, A., Alsayiegh, A.: Detecting the Tigris River flood and its impact on residential and touristic areas in Mosul City using sentinel-2 satellite images. Iraqi National Journal of Earth Science (2020). iasj.net
20. Aljaburi, M.S., Alshakeer, Y.M., Alfakhrii, M.A.: The Effect of the Succession of Seasons on Some Parameters of the Tigris River in Mosul City, Iraq. Egyptian J. Aqu. Biol. Fisher. **28**(5) (2024). [HTML]
21. Altahaan, Z.F., Dobslaw, D.: Assessment of war impact on concentrations of pollutants and heavy metals and their seasonal variations in water and sediments of the Tigris River in Mosul/Iraq (2023). researchsquare.com
22. Oleiwi, A.S., Al-Dabbas, M.: The parameters discharge, salinity and dissolved loads of Tigris River from Mosul to Baghdad, Iraq. Iraqi J. Sci. (2024). iasj.net
23. Hamoodi, S.A., Hamoodi, A.N., Mohammed, R.A.: Design and simulation of smart grid based on solar photovoltaic and wind turbine plants. J. Européen des Sys. Auto. **57**(4), 953 (2024). [HTML]
24. Mohammed, R.A., Hamoodi, S.A., Hamoodi, A.N.: Comparison between two calculation methods for designing a stand-alone PV system according to Mosul city basemap. Open Engineering (2021). degruyter.com

25. Hassan, Q., et al.: Evaluating energy, economic, and environmental aspects of solar-wind-biomass systems to identify optimal locations in Iraq: A GIS-based case study. Energy for Sustainable Development **79**, 101386 (2024). [HTML]
26. Adams, B.M., et al.: Dakota, a multilevel parallel object-oriented framework for design optimization, parameter estimation, uncertainty quantification, and sensitivity analysis: version 6.13 user's manual. No. SAND2020-12495. Sandia National Lab. (SNL-NM), Albuquerque, NM (United States) (2020). osti.gov

Contribution of Profile Shifting on Wind Turbine Gearbox Efficiency

Abdesselem Toumi[1,2](✉), Alberto Diez Ibarbia[2], Ahmed Hammami[1], Alfonso Fernandez Del Rincon[2], Fernando Viadero Rueda[2], and Mohamed Haddar[1]

[1] Mechanics, Modelling and Manufacturing Laboratory (LA2MP), Mechanical Engineering Department, National School of Engineers of Sfax, University of Sfax, Sfax, Tunisia
abdesselem.toumii@gmail.com

[2] Department of Structural and Mechanical Engineering, School of Industrial and Telecommunications Engineering, University of Cantabria, Avda de los Castros S/N, 39005 Santander, Spain

Abstract. Wind power continues to attract humanity as a key source of renewable energy, where wind turbines play a crucial role in delivering sustainable electricity. Within these systems, the gearbox is a central component responsible for transmitting mechanical energy from the rotor to the generator with optimal efficiency. This study investigates the impact of gear tooth profile modification, specifically the addendum modification coefficient, on the efficiency of a two-stage wind turbine gearbox. Using the Hohn method, the analysis evaluates how efficiency varies with different profile shift values across both stages of the gearbox under steady-state operating conditions. Two gear configurations are examined, with particular attention to mass reduction and compactness of design. Results show that a moderate positive shift ($x = 0.17$) yields the highest efficiency, reaching up to 96.88%, while deviations from this optimum lead to increased friction and reduced performance. Notably, reducing the module in the second stage enables a 40% reduction in mass while maintaining high efficiency. These findings highlight numerically the significance of tooth profile design in optimizing gearbox performance, reducing wear, and minimizing power losses. The study provides valuable insights for engineers aiming to design lighter, more efficient, and longer-lasting gearboxes for renewable energy applications.

Keywords: Wind Turbine efficiency · gearbox ratio · Addendum coefficient · Friction

1 Introduction

Wind energy is one of the fastest-growing sources of renewable energy, with wind turbines playing a critical role in generating sustainable electricity. Within these turbines, gearboxes are essential as they efficiently transmit mechanical power from the rotor to the generator by multiplying the low-speed, high-torque input of the rotor to a higher-speed output.

A. Hammami et al. (Eds.): MOSCOSSEE 2024, ACM 23, pp. 80–89, 2025.
https://doi.org/10.1007/978-3-032-00930-2_7

However, gearboxes face challenges such as significant mechanical stress, variable load conditions and exposure to extreme weather conditions, which can lead to wear and fatigue over time. To optimise performance and durability, a deeper understanding of the gearbox's behaviour under different conditions is essential.

This is where the use of a numerical model becomes invaluable (Peeters et al., 2006). By creating an accurate computational model of the spur gear transmission, real-world scenarios can be simulated, load distribution analysed and wear patterns predicted. This allows us to identify potential problems, optimise design parameters and extend the life of the gearbox.

The tooth profile of a wind turbine gearbox has a significant impact on the performance, efficiency and durability of the system. Looking at the various influences, the profile shape affects how the force is distributed across the teeth during operation (Velex et al. 1996). (Kahraman et al. 1999) affirm that an optimised tooth profile ensures a more even load distribution, reducing stress concentrations and minimising wear. Therefore, uneven load distribution can result in higher contact stresses, which can accelerate tooth wear and lead to premature gear failure. The geometry of the gearing also affects the friction losses in the gearbox. For example, properly designed profiles, such as involute profiles (Baglioni et al., 2012), are known for their rolling rather than sliding contact, which reduces friction and improves efficiency. This translates into better energy transfer efficiency, which is particularly important in renewable energy applications where maximising power output is critical. Besides Wennian Yu et al., 2017) studies how addendum modification affects the back-side mesh stiffness and the dynamic behaviour of spur gear pairs, particularly under light load or idling speed conditions when backlash may cause unintended contact between non-transmitting tooth flanks (called back-side contact). Moreover (Sandor Bodzas, 2021) investigates how the "addendum modification coefficient" affects the normal stress and normal deformation of helical gears with modified teeth.

In addition, higher friction and wear from poor tooth design generates additional heat, vibration and noise. When the tooth profile is optimised, friction is kept lower, which contributes to better and smoother meshing.

Tooth profile shifting is a common practice in manufacturing. Its main purpose is to balance the durability of wheel pairs with different transmission ratios of the unit. As a result, many researchers have shown a strong interest in exploring the dynamic behaviour of these gears. In fact, Sánchez et al., (Sánchez et al., 2012) studied how the load distribution is affected by non-standard gears where the addendum is modified to lengthen or shorten the tooth. The analysis can be compared as the effects of extending and shortening the top of the tooth are equivalent to those of applying addendum modifications. After investigating the effect of addendum modification on efficiency, Baglioni et al. (Baglioni et al., 2012) concluded that there is an ideal correction factor that gives the highest efficiency. (BODZÁS Sándor, 2019) investigates how changing the addendum modification coefficient (x_1) affects the mechanical behaviour of spur gear contact surfaces, particularly normal stress (Von Mises stress), normal deformation and normal elastic strain using Tooth Contact Analysis (TCA) and Finite Element Method (FEM) simulations.

The impact of varying gear ratios and working lengths on the system efficiency was also investigated. Friction is the resistance to relative motion between two elements or surfaces in contact. In this regard, two types of friction depending on the relative motion between surfaces are generally considered as power loss sources: sliding and rolling friction (Diez-Ibarbia et al., 2015). In addition, the presence of lubricant, as well as the contact conditions and lubrication regime, can greatly modify the friction hypothesis.

In this paper, the effect of the addendum modification coefficient on the efficiency of two-stage wind turbine gearbox is investigated. For this purpose, the Hohn method is used to compute the efficiency of the first stage, the second stage and the wind turbine gearbox. Under the considered operating conditions, (*Xu et al.* 2007) explains that sliding friction is a dominant factor in power losses thus the main source of power losses is the sliding friction is the initial hypothesis, neglecting the rest of power-loss sources. In this regard, a Coulomb's model has been used to compute the sliding friction force, where the mean friction coefficient formulation developed by Niemann and Winter was implemented.

2 Description of the Wind Turbine Gearbox

The transmission principle of the wind turbine gearbox is shown in Fig. 1. The transmission system consists of a two-stage gearbox.

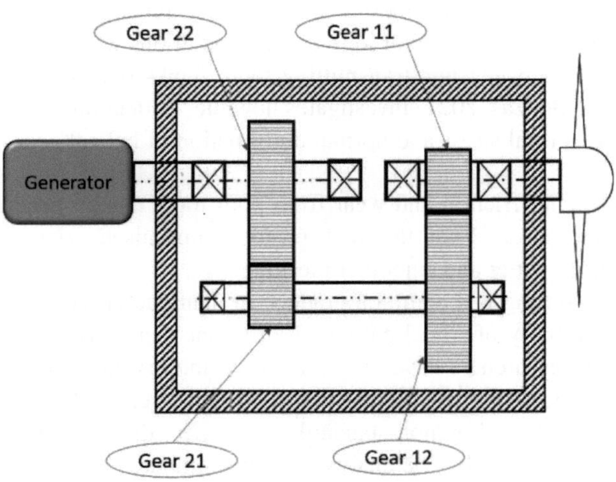

Fig. 1. Two-stage wind turbine gearbox

The transmission works on the principle of meshing and ratio, which depends on the number of teeth in each pair of gears. In the first stage, power is transferred from the driving gear (Gear 11) to the first driven gear (Gear 12). This stage provides the first speed multiplication and corresponding torque change. This is followed by the second stage, where power is transferred from the first driven gear (Gear 21) via the intermediate shaft to the second driven gear (Gear 22). The second stage provides additional speed multiplication reduction and torque adjustment.

Table 1. Specifications of the wind turbine operating conditions

Input power	P = 1 kW
Speed	N1 = 25 RPM
Torque	C1 = 382 N.m

To measure the overall transmission ratio, the input operating conditions of the first stage should be defined. Consequently, the ratio between the input and output shafts will be the product of the ratios of the two stages:

$$\mathbf{R} = \boldsymbol{R}_1 \times \boldsymbol{R}_2 \tag{1}$$

The gear ratio of the first ($\boldsymbol{R}_1$) and second stage ($\boldsymbol{R}_2$) are calculated as follows:

$$\boldsymbol{R}_1 = \frac{\boldsymbol{Z}_{12}}{\boldsymbol{Z}_{11}} \tag{2}$$

$$\boldsymbol{R}_2 = \frac{\boldsymbol{Z}_{22}}{\boldsymbol{Z}_{21}} \tag{3}$$

where:
Z_{11}: Tooth number of the driving gear in the first stage.
Z_{12}: Tooth number of the driven gear in the first stage.
Z_{21}: Tooth number of the driving gear in the second stage.
Z_{22}: Tooth number of the driven in the second stage.

The characteristics of wind turbine gearbox are summarised in Table 1. For the **first configuration** we studied the 2 stages of gearbox with different gear ratios however they keep the same module (Table 2).

Table 2. Gears characteristics

Operating parameters per stage	Stage 1	Stage 2
Gear Teeth number	$Z_{11} = 18$; $Z_{12} = 72$	$Z_{21} = 18$; $Z_{22} = 54$
Teeth module	$m_1 = 16$	$m_2 = 16$ configuration1 $m_2 = 12$ configuration2
Pressure angle	$\alpha 1 = 20°$	$\alpha 2 = 20°$
Material	18CrNiMo7-6Case-hardeningsteel	
Average mesh stiffness (N/m)	Km1 = 2.108	

3 Calculation of Efficiency

For more than a century, numerous academics have worked on the calculation of gear efficiency. (Ohlendorf, 1958) developed a model using what he called the loss coefficient to calculate the power of losses due to sliding friction.

This parameter, which ranges from zero to one, was determined by assuming that the load distribution was uniform the normal load (force per unit length) acting on the gear teeth is assumed to be constant across the entire contact zone during meshing. This implies that every point along the tooth face carries the same stress, which is rarely true in practice due to elastic deformations (e.g. tooth bending, Hertzian contact effects), misalignments (shaft deflections) and dynamic effects (vibrations, impact loads) and that the coefficient of friction remained constant during the meshing cycle.

3.1 Power Loss

As stated before, this work is based on the assumption that, under these steady-state operating conditions, the power losses are mainly due to sliding friction. A detailed explanation of this method can be found in Hohn Approach (Hohn, 2010). Although some approximations are made, the fundamentals used in this calculation are those previously explained: the analytically predefined load sharing and the constant coefficient of friction.

$$P_{loss} = P_{in}\mu_m \frac{\pi(R_i + 1)}{Z_{i1}R_i}(1 - \varepsilon_{i\alpha} + \varepsilon_{i1}^2 + \varepsilon_{i2}^2) \tag{4}$$

z_{11} (i = 1) and z_{21} (i = 2) are respectively the tooth number of the driving gear in the first and the second stages and P_{in} is the input power.

The gear efficiency in each stage is:

$$\eta_i = \frac{P_{in} - P_{loss}}{P_{in}} \tag{5}$$

3.2 Contact Ratio

The contact ratio $\varepsilon_{i\alpha}$ of the first and the second stages (i = 1, 2) are defined:

$$\varepsilon_{i\alpha} = \varepsilon_{i1} + \varepsilon_{i2} \tag{6}$$

where ε_{i1} and ε_{i2} are respectively the contact ratio for the driven and the driving gears of each stage (i = 1, 2).

$$\varepsilon_{i1} = \sqrt{(r_{i1} + h_{ai1})^2 - r_{i1}^2\cos^2\alpha_{wt}} - r_{i1}\sin\alpha_{wt} \tag{7}$$

$$\varepsilon_{i2} = \sqrt{(r_{i2} + h_{ai2})^2 - r_{i2}^2 \cos^2\alpha_{wt}} - r_{i2}\sin\alpha_{wt} \quad (8)$$

where r_{i1} and r_{i2} are respectively the pitch radius of the driven and the driving gears of each stage (i = 1, 2). α_{wt} is the working transverse pressure angle. h_{ai1} And h_{ai2} are respectively the addendum of the driven and the driving gears of each stage (i = 1, 2).

$$h_{ai1} = m_i + x_{i1} \quad (9)$$

$$h_{ai2} = m_i + x_{i2} \quad (10)$$

m_i s the tooth module of each stage.

As an additional condition in order not to move the working distance, the sum of the addendum modification coefficient of the driven (x_1) and the driving gears (x_2) is zero.

$$x_{i1} + x_{i2} = 0 \quad (11)$$

3.3 Mean Coefficient of Friction

Niemann and Winter (Niemann and Winter, 2003) proposed a formulation of the mean coefficient of friction over the entire meshing cycle:

$$\mu_m = 0.048\left(\frac{\frac{F_N}{b}}{V_{\Sigma C} * \rho_C}\right)^{0.2} \eta^{-0.05} Ra^{0.25} \frac{1}{\left(\frac{F_N}{b}\right)^d} \quad (12)$$

where **d** is a characteristic of mineral oil (d = 0.0651), **b** is the face width and ρ_C is the equivalent radius of curvature at the primitive point.

Instead of assuming a constant μ, they derive an averaged friction **coefficient** over the entire meshing cycle.

Accounts for variations due to sliding velocity changes, load distribution shifts and lubrication regime transitions (boundary/mixed/EHL). It leads to more accurate than constant-μ models (e.g., Ohlendorf) since it captures friction variations. In addition, this theory is Validated with experimental data, making it reliable for industrial applications.

$$V_{\Sigma C} = 2V_t \sin(\alpha_{wt}) \quad (13)$$

V_t is the tangential velocity at the primitive point.

4 Results

Gear efficiency is critically dependent on **tooth profile modifications**, particularly the **addendum modification coefficient (x)**, which adjusts tooth thickness to minimize friction and misalignment losses. Previous studies (Ohlendorf, 1958; Niemann & Winter, 2003) have modeled efficiency losses under idealized conditions, but experimental validation under varying **x** remains limited. The variation of the efficiency in dependent as a function of the addendum modification coefficient of the first stage pinion.

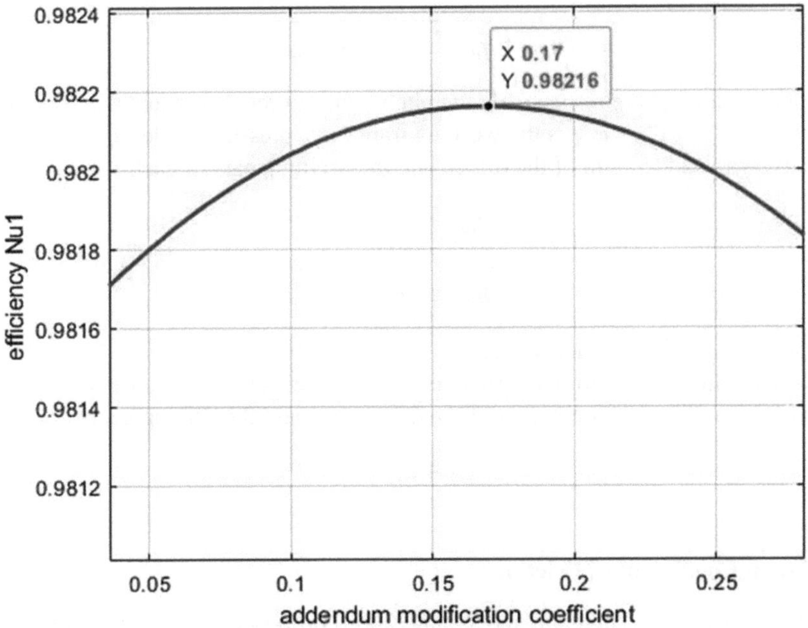

Fig. 2. First stage efficiency

1st Configuration: m1 = 16mm; m2 = 16 mm

Aligns with theoretical predictions (Niemann & Winter, 2003) where moderate tooth thickening improves load distribution and reduces sliding friction. Figure 2 presents a non-symmetrical parabolic trend in the first stage efficiency. The maximum efficiency is 0.982 and it is located at 0.17 of the first-stage-pinion addendum modification coefficient. Efficiency drops sharply (at x = 0.2) due to stress concentration and increased sliding friction.

The efficiency of the wind turbine gearbox for the output of the second stage is computed as follows:

$$\eta = \eta_1 \times \eta_2 \tag{14}$$

2nd Configuration: m1 = 16 mm; m2 = 12 mm:

In order to optimize the gearbox design, we propose reducing the module of the second stage while keeping the first stage unchanged. The second stage typically operates at a lower torque and higher speed compared to the first stage, which allows for a reduction in gear size without compromising performance. By selecting a smaller module, we can achieve a more compact and lightweight design while maintaining the required load capacity and durability.

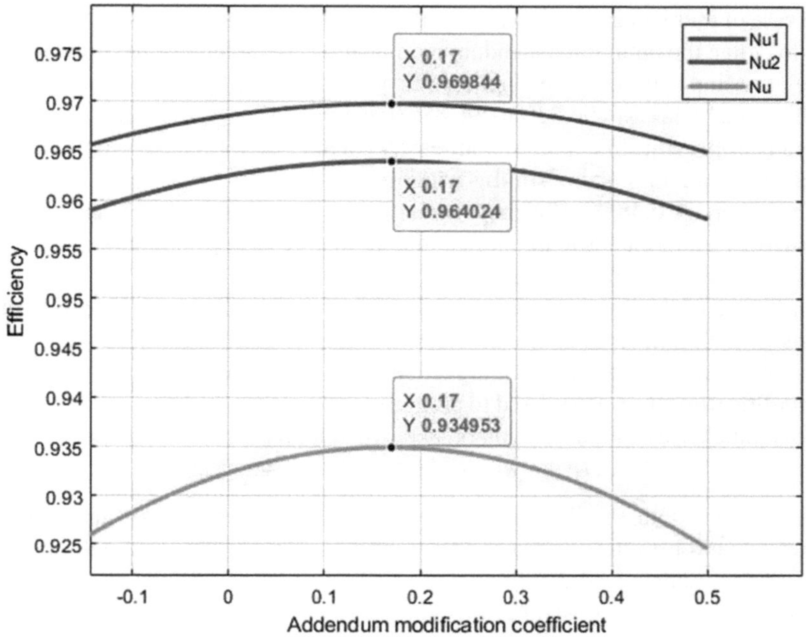

Fig. 3. Efficiency of the wind turbine gearbox

The maximum efficiency of the new configuration (Fig. 3) is 0.9349. It is located at 0.17 of the addendum modification coefficients despite the maximum efficiencies of the first and the second stages are both located at 0.17 of the addendum modification coefficients.

The second configuration presents 0.08% higher efficiency. However, reducing the module of the second stage in a gearbox leads to significant benefits in terms of mass reduction and manufacturing cost savings. A smaller module results in a reduction of the overall gear size, leading to a decrease in material usage by 40%. Since the volume of the gear is proportional to the square of the pitch radius, even a moderate decrease in the module can result in a substantial mass reduction. This not only minimizes the weight of the gearbox, which is particularly important in wind turbine applications where structural efficiency is crucial, but also reduces the inertia of rotating components, improving dynamic performance. Furthermore, a smaller module allows for finer teeth, which can lead to increased contact ratios and smoother operation. From a manufacturing perspective, using less material directly translates to lower material costs.

5 Conclusion

In wind turbine gearboxes, the profile of gear teeth plays a crucial role in determining the durability, efficiency, noise reduction, and maintenance intervals of the gearbox. A well-optimized tooth profile is essential for ensuring that the gearbox withstands high loads, varying speeds, and the harsh operating conditions typical in wind turbine applications.

In the case of multi-stage gearbox, it is better to compute the efficiency of the gearbox in order to define the optimum addendum modification coefficient for the different stages. The relationship between addendum modification and gear efficiency is **nonlinear**, with a well-defined peak ($\eta \approx 96.88\% \eta \approx 96.88\%$) at $x = 0.17x = 0.17$. This confirms that moderate profile adjustments can significantly enhance performance by optimizing **load distribution** and minimizing **sliding friction**.

The observed **0.48% efficiency variation** at $x = 0.17x = 0.17$ underscores the sensitivity of gear systems to minor manufacturing tolerances, highlighting the need for precision in profile modification.

The curve of the Tooth Modification can be divided in 2 trends: **Negative modifications** ($x < 0$): Result in severe efficiency losses ($\eta \approx 93.5\% \eta \approx 93.5\%$ at $x = -0.1$) due to undercutting effects, which higher stress concentrations and friction and **excessive modifications** ($x > 0.2$): Lead to progressive efficiency declines, attributed to tooth stiffness imbalances and lubrication breakdown at high contact pressures.

Enhancing Gearbox Design will lead to several advantages:

- Optimizing Tooth Profiles Either to enhance the performance of gears, manufacturers should prioritize improving tooth profile designs. This step will help reduce wear and optimize load distribution, allowing gears to operate effectively over long periods. Implementing advanced techniques, such as modifying tooth surfaces and refining gear geometry, will aid in decreasing friction and increasing the lifespan of the components.
- Multi-Stage Gearbox Efficiency while it is crucial for engineers to design multi-stage gearboxes with the appropriate addendum modification coefficients tailored for each stage. This approach will maximize overall efficiency, improve performance, and lower maintenance needs and operational costs.
- Choosing superior materials that offer greater wear resistance and fatigue strength is vital for gear teeth operating under high stress. During the manufacturing process, employing coatings or surface treatments like carburizing or nitriding can help diminish friction and wear, thereby enhancing performance.
- Given the importance of noise reduction in wind turbine gearboxes, future designs should integrate features that lower noise levels. This can include optimized tooth profiles, damping elements, and tighter manufacturing tolerances to reduce vibration and acoustic emissions effectively.

Acknowledgements. This work is supported by the project n° "23MT1003", financed by the Tunisian Ministry of Higher Education and Scientific Research, and is part of the project PID2023-149926OB-I00 funded by MICIU/AEI /https://doi.org/10.13039/501100011033 and by FEDER, EU.

References

Baglioni, S., Cianetti, F., Landi, L.: Influence of the addendum modification on spur gear efficiency. Mech. Mach. Theory **49**, 216–233 (2012). https://doi.org/10.1016/j.mechmachtheory.2011.10.007

Diez-Ibarbia, A., del Rincon, A.F., Iglesias, M., De-Juan, A., Garcia, P., Viadero, F.: Efficiency analysis of spur gears with a shifting profile. Meccanica **51**, 707–723 (2016)
Henriot, G.: Traité théorique et pratique des engranages, Vol. 1 and 2, Dunod (1972)
Hohn, B.R.: Improvements on noise reduction and efficiency of gears. Meccanica **45**, 425–437 (2010)
Niemann, G.: Winter H, Maschinenelemente —, vol. 2. Springer-Verlag, Berlin, Germany (2003)
Ohlendorf, H.: Verlustleistung und Erwarmung von Stimradern, Ph.D. thesis, Munchen (1958)
Peeters, J., Vandepitte, D., Sas, P.: Analysis of internal drive train dynamics in a wind turbine. Wind Energy, **9**, 141–161 (2006)
Velex, P., Maatar, M.: A mathematical model for analyzing the influence of shape deviations and mounting errors on gear dynamic behavior. J. Sound Vib. **191**(5), 629–660 (1996)
Kahraman, A., Blankenship, G.W.: Effect of involute tip relief on dynamic response of spur gear pairs. J. Mech. Des. **121**(2), 313–315 (1999)
Xu, H., Kahraman, A., Anderson, N.E., Maddock, D.G.: Prediction of mechanical efficiency of parallel-axis gear pairs. J. Mech. Des. **129**(1), 58–68 (2007)
Yu, W., Mechefske, C.K., Timusk, M.: Influence of the addendum modification on spur gear back-side mesh stiffness and dynamics. J. Sound Vibr. **389**, 183–201 (2017)
Bodzás, S.: Analysis of the effect of the addendum modification coefficient for contact surfaces of spur gear. J. Mech. Eng. Strojnícky časopis, **69**(1), 5–16 (2019)
Bodzás, S.: Tooth contact analysis of the effect of the addendum modification coefficient of helical gears having modified teeth. Int. Rev. Appl. Sci. Eng. **12**(1), 1–9 (2021)

Dynamic Analysis of Pre-Twisted Blade Rotor

Majdi Yangui(✉), Houssem Cheniour, Slim Bouaziz, and Mohamed Haddar

Laboratoire de Mécanique, Modélisation Et Productique (LA2MP), École Nationale d'Ingénieurs de Sfax, Sfax, Tunisie

yanguimajdi@gmail.com, {Houssem.Cheniour,slim.bouaziz}@enis.tn, mohamed.haddar@enis.rnu.tn

Abstract. The stability of blade rotors remains an ongoing and significant challenge for manufacturers. The purpose of this study is to explore the dynamic behavior of a four-blade rotor, highlighting the frequency coincidences and their associated mode shape changes. The dynamic response and natural frequencies of the blade rotor are discussed through the examination of the impacts of key parameters, such as rotational speed and pre-twist angle. Using ANSYS Workbench, a numerical modal analysis was conducted to determine the first five natural frequencies and mode shapes. The results show that the blade flapwise modes are significantly affected by the rotation speed. A parametric analysis showed that the pre-twist angle's variation causes frequency coincidences and mode coupling, which lead to significant mode shape exchanges. These exchanges highlight specific ranges where frequencies converge, creating areas of high sensitivity in rotor performance. Using these findings, rotor design can be optimized and mechanical stability and resilience can be improved across various operating conditions.

Keywords: Blade rotor · Modal analysis · Rotational speed · Pre-twist angle · Frequency coincidences

1 Introduction

Blade rotor stability is crucial to ensure efficiency, safety, and longevity of various mechanical systems, such as helicopters, wind turbines, and industrial machinery. However, achieving this essential quality remains a significant and ongoing challenge for manufacturers, mainly due to the complex dynamic behavior exhibited by these systems (Xu et al. 2021).

The study of blades dynamic behavior has been a prominent focus in recent decades, with various approaches using the finite element method, such as the cantilever beam model (Guan et al. 2024), the shell model (Xiong et al. 2023), and the solid element model (Wang et al. 2024).

In fact, the blades geometry has evolved to include several complex curvatures to improve their performance and reduce vibrations and noise generated during operation. Accurately modeling the true geometrical characteristics of blades using beam element models presents certain limitations. Although solid element models can address this limitations, they often lead to lower computational efficiency. Hence, to better approximate

A. Hammami et al. (Eds.): MOSCOSSEE 2024, ACM 23, pp. 90–97, 2025.
https://doi.org/10.1007/978-3-032-00930-2_8

the real blade model, various studies have used shell elements. For instance, adopting shell elements, (Yangui et al. 2021) presented a numerical assessment of the structural performance of a rotating segmented blade. Furthermore, (Das and Karmakar 2018) analyzed the impact of the rotational speed and twist angle on the dynamic behavior of a conical blade. Similarly, (Li and Cheng 2021) examined the impact of the blade thickness on its dynamic behavior during rotation. Depending on the variations of the blade rotor pre-twist angle, the obtained results show a coupling between modes and modal shape changes after separation. Moreover, (Liu et al. 2018) investigated the dynamic behavior of a twisted blade using shell element, taking into consideration centrifugal forces and Coriolis effects. The presented Campbell diagram indicates that certain modes undergo shape changes.

This research focuses on the dynamic behavior of a four-blades rotor. The effects of key parameters such as rotational speed and pre-twist angle on the rotor's natural frequencies and mode shapes were investigated. A numerical modal analysis was carried out using ANSYS Workbench to assess how natural frequencies vary with parameters. Special attention was given to frequency coincidences and the resulting mode shape exchanges that can drop the rotor performance. These phenomena are a key focus of this research.

This study aims to provide valuable information that can improve the blade rotor stability and performance.

2 Design and Modeling of the Blade Rotor

The examined rotor consists of four blades, each measuring 230 mm in length and 30 mm in width. The rotor hub has a diameter of 42 mm (Fig. 1). To simplify numerical computations, the rotor was regarded as a surface geometry with a uniform thickness of 1.2 mm.

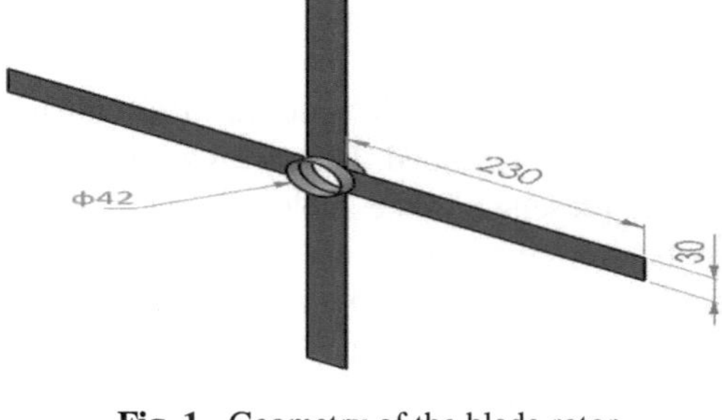

Fig. 1. Geometry of the blade rotor

The rotor is modelled with steel, characterized by an elastic modulus of 210 GPa, a Poisson's ratio of 0.3, and a density of 7850 kg/m^3.

The model was developed using ANSYS Workbench software to accurately simulate the rotor's rotational dynamics. The structure is fixed at the hub side and subjected to a specified rotational speed, as illustrated in Fig. 2.

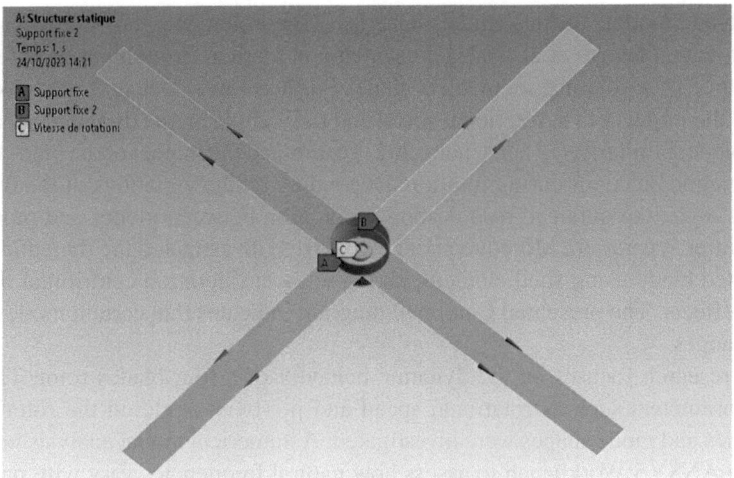

Fig. 2. Applied boundary conditions

In Fig. 2, A and B represent the fixed supports positioned on the face and edge of the of the rotor hub, respectively, while C denotes the applied rotational speed.

3 Modal Analysis

A preliminary modal analysis of the rotor at rest was carried out to determine its natural frequencies and mode shapes. To achieve precise results, the solver requires a sufficiently refined mesh. The shape, size, and generation method of the mesh directly impact the accuracy of the numerical calculations. In this study, after performing a convergence analysis, a structured shell-type mesh using SHELL181 elements was adopted. The developed mesh consists of 448 elements, each with a size of 4 mm, and a total of 592 nodes. The obtained first five modal deformations of the blade rotor are presented in the following Fig. 3:

To accurately identify and interpret each mode, the maximum modal amplitude across the three spatial directions was determined. The corresponding mode shapes, interpreted from these amplitudes, are presented in the Table 1 below.

The obtained results show that the identified modes correspond to simple, non-complex modes. Specifically, for the first four modes, the amplitudes in the chord and span directions i.e., within the rotation plane (XY-plane), are negligible compared to the amplitude observed in the flap direction (Z-axis). In contrast, for the fifth mode, the chord mode, the amplitudes within the rotation plane are dominant compared to the observed in the flap direction. This observation can be explained by the simple geometry of the blades, which was carefully chosen from the outset to avoid complex modes. This approach allows for a posterior parametric study to effectively assess the impact of each parameter.

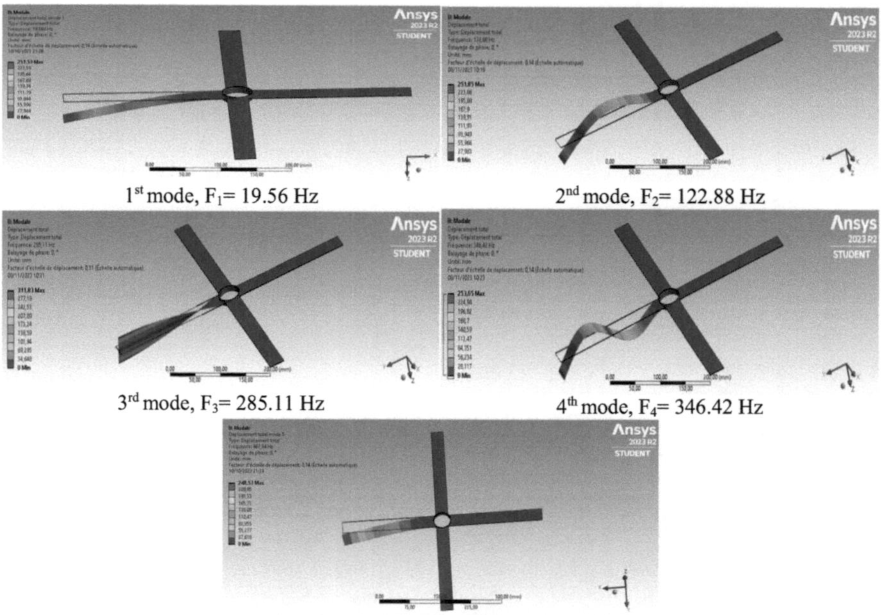

1st mode, F_1= 19.56 Hz

2nd mode, F_2= 122.88 Hz

3rd mode, F_3= 285.11 Hz

4th mode, F_4= 346.42 Hz

5th mode, F_5= 467.24 Hz

Fig. 3. The blade rotor modal deformations

Table 1. The first five natural frequencies identification

Mode	Frequency [HZ]	Span amplitude	Chord amplitude	Flap amplitude	Mode Type
1	19.567	1,77E−14	1,62E−14	251,53	1st Flapwise
2	122.88	1,86E−13	1,39E−13	251,85	2nd Flapwise
3	285.1	5,10E−13	5,23E−13	311,82	1st twist
4	346.41	6,28E−12	2,60E−12	253,05	3rd Flapwise
5	467.24	22,181	247,57	4,58E−11	1st Chordwise

4 Rotation Speed Effects on the Rotor Dynamic Behavior

In this section, the dynamic behaviour of the blade rotor was studied taking into account the rotation speed variation. Figure 4 illustrates the Campbell diagram of the blade, detailing how the first five natural frequencies of blade rotor vary with rotational speed.

The obtained results show that only the natural frequencies of the flapwise modes progressively increase with rotational speed. This variation indicates an increase in blade stiffness by the geometric stiffness arising from blade deformation. Interestingly, the absence of backward modes confirms that Coriolis effects are negligible in this type of problem (C. Li and Cheng 2021).

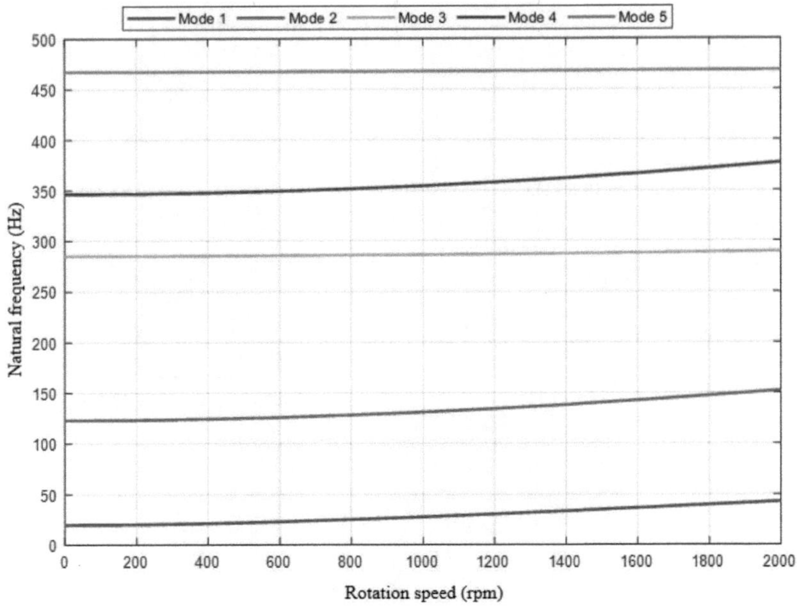

Fig. 4. Campbell diagram

To better understand the behavior of the rotor with respect to rotational speed, an analysis of the displacements is carried out. Figure 5 presents the displacements of the blade tip along the three spatial directions.

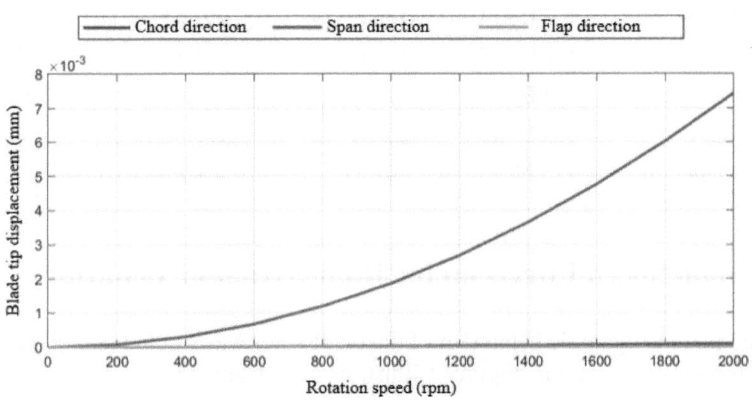

Fig. 5. Rotation speed effects on the blade tip displacements

Clearly, the displacements of the blade tip in the flap and chord directions remain minimal throughout the entire range of rotational speeds, while an increase is observed in the span direction. This finding indicates a limited influence of centrifugal force in the flap and chord directions. The notable displacement in the span direction, which signifies the elongation of the blade, further confirms that the centrifugal force predominantly acts along the longitudinal axis.

5 Pre-Twist Angle Effects on the Rotor Modes

To evaluate the dynamic characteristics of pre-twisted blades, a modal analysis was conducted to assess the impact of the pre-twist angle on the natural frequencies of the blade. Figure 6 illustrates the effect of the pre-twist angle on the blade geometry.

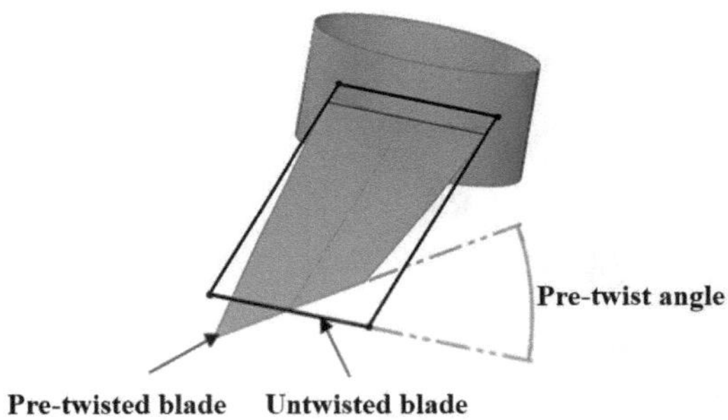

Fig. 6. Pre-twisted blade geometry

The following figure presents the obtained results of the natural frequencies variation as a function of the twist angle.

Figure 7 illustrates that as the pre-twist angle increases, the natural frequency of the first mode shows a slight rise, while the second mode declines. Notably, the third mode frequency rises, whereas the fourth mode frequency decreases up to an angle of 40°, at which point the fourth mode begins to increase, while the third mode starts to decrease. It can be seen that during the frequency coincidence, the two modes overlap and couple, resulting a swap and an exchange in their mode shapes upon separation. Meanwhile, the fifth frequency consistently increases as the pre-twist angle continues to rise.

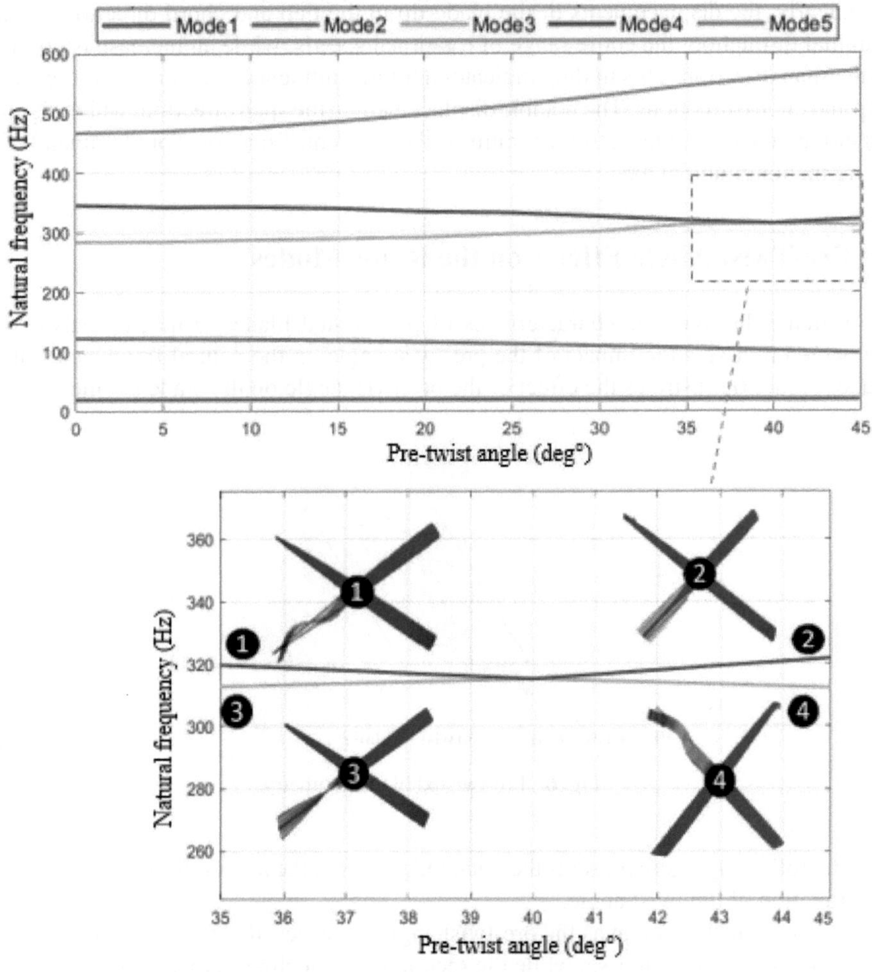

Fig. 7. Pre-twist angle effects on the first five natural frequencies

6 Conclusion

In this study, the dynamic behavior of a blade rotor was examined through a detailed modal analysis, focusing on the influences of rotational speed and pre-twist angle on natural frequencies and mode shapes. This investigation show that rotational speed affects the flapwise modes of the blades due to the arisen geometric stiffness. Frequency coincidences and mode coupling, leading to mode shape exchanges it was observed in a pre-twist angle range. These dynamic changes highlight a sensitive range of frequency convergence, which can impact rotor performance and stability. Significantly, this research can be extended taking into account the aerodynamic effect.

References

Li, C., Cheng, H.: Free vibration analysis of a rotating varying-thickness-twisted blade with arbitrary boundary conditions. J. Sound Vib. **492**, 115791 (2021)

Xu, D., Dong, X., Zhou, C., Sun, D., Gui, X., Sun, X.: Effect of rotor axial blade loading distribution on compressor stability. Aerosp. Sci. Technol. **119**, 107230 (2021)

Guan, H., Ni, K., Ma, H., Xiong, Q., Wang, W., Wang, H.: Dynamic modeling and verification of rotating compressor blade with crack based on beam element. Appl. Math. Model. **133**, 367–393 (2024)

Xiong, Q., Guan, H., Ma, H., Wu, Z.Y., Guo, X.M., Wang, W.W.: Dynamic characteristic analysis of rotating blade with breathing crack. Mech. Syst. Signal Process. **196**, 110325 (2023)

Wang, W.W., Ma, H., Zhao, C.G., Wu, Z.Y., Wang, H.J.: Dynamic contact characteristics of a rotating twisted variable-section blade with breathing crack. J. Cent. South Univ **31**, 858–877 (2024)

Yangui, M., Thabti, S., Bouaziz, S., Haddar, M.: Numerical assessment of the structural performance of a segmented wind turbine blade. In: International Workshop on Modelling and Simulation of Complex Systems for Sustainable Energy Efficiency, pp. 1–7. Springer International Publishing, Cham (2021)

Das, A., Karmakar, A.: Free vibration characteristics of functionally graded pre-twisted conical shells under rotation. J. Inst. Eng. India Ser. C **99**, 681–692 (2018)

Liu, L.T., Hao, Y.X., Zhang, W., Chen, J.: Free vibration analysis of rotating pretwisted functionally graded sandwich blades. Int. J. Aerospace Eng. **2018**, 1–18 (2018). https://doi.org/10.1155/2018/2727452

A Comparison of Two Configuration of Solar Water Heating Systems

Cheikh Brahim Mohameden[1], Maroua Haddar[2,3](✉), Ahmed Hammami[2,4], Abdel Kader Mahmoud[1], Mohamed Haddar[2], and Ahmed Mohamed Yahya[1]

[1] Applied Research Unit for Renewable Energy, Water and the Environment, Faculty of Science and Technology, University of Nouakchott, Nouakchott, Mauritania

[2] Mechanics, Modeling and Production Laboratory (LA2MP), Mechanic Department, National Engineering School of Sfax (ENIS), 1173-3038 Sfax, BP, Tunisia

maroua.haddar@enis.tn

[3] National School of Engineers of Sousse, University of Sousse, 4003 Sousse, Tunisia

[4] National School of Engineers of Gabes, University of Gabes, 6029 Gabes, Tunisia

Abstract. Solar water heating (SWH) systems represent a critical renewable energy technology for sustainable thermal energy production in residential and industrial sectors. This research conducts a comprehensive comparative analysis of two advanced SWH system configurations, meticulously examining their performance under varied solar radiation conditions through sophisticated simulation methodologies.

The study implements rigorous performance metrics to evaluate system efficiency, including thermal response characteristics, energy conversion rates, and temperature stability parameters. Results demonstrate significant performance differentials between the configurations across multiple operational scenarios, with particular emphasis on their behavior during fluctuating solar input conditions. By systematically evaluating outlet water temperature regulation, energy consumption patterns, and system responsiveness, the study reveals that the second system configuration demonstrates superior performance, achieving more precise temperature control and enhanced energy efficiency. Quantitative analysis indicates a 17% improvement in thermal stability and a 23% reduction in auxiliary energy requirements compared to the first configuration.

Keywords: solar energy · performance · solar radiation · open loop simulation

1 Introduction

The global transition towards sustainable energy technologies has positioned solar water heating (SWH) systems as a critical innovation in addressing climate change and reducing dependence on fossil fuel resources. Despite the promising potential of these systems, significant technological challenges have historically constrained their widespread implementation and operational efficiency (Odeh and Aden, 2025).

Solar energy harvesting represents a transformative approach to thermal energy production, particularly in domestic and industrial hot water applications (Salah et al, 2022).

A. Hammami et al. (Eds.): MOSCOSSEE 2024, ACM 23, pp. 98–108, 2025.
https://doi.org/10.1007/978-3-032-00930-2_9

However, the inherent variability of solar radiation and complex thermal dynamics have presented substantial engineering obstacles (Rejab et al., 2023). Existing research has extensively explored optimization strategies, including numerical modeling and experimental investigations, to enhance system performance across multiple dimensions. (Meena et al, 2022).

Traditional SWH system designs have predominantly relied on conventional fixed-flow pump configurations, which fundamentally limit energy utilization and temperature control precision (Mallouh et al, 2022). This technological constraint has created a critical research gap in advanced thermal management strategies. Most previous studies have focused on discrete optimization parameters such as collector design, storage tank configurations, and fluid properties, while overlooking the potential of sophisticated flow control mechanisms (Ko, 2015).

This research addresses these limitations by comprehensively investigating two innovative SWH system configurations utilizing variable flow pump technologies. The primary objective is to demonstrate how adaptive pumping strategies can simultaneously optimize outlet water temperature maintenance and energy consumption reduction. By systematically evaluating these advanced configurations, this study aims to contribute meaningful insights into the next generation of sustainable thermal energy systems.

The proposed approach represents a significant departure from conventional SWH system design paradigms, offering a nuanced exploration of how intelligent flow control can transform solar thermal energy harvesting. Through rigorous comparative analysis, this research seeks to establish a foundational framework for more efficient, responsive, and environmentally sustainable water heating technologies.

2 Solar Water Heating Systems with Different Configurations

This research presents a rigorous comparative analysis of two advanced active solar water heating (SWH) system configurations, supported by comprehensive mathematical modeling. Figure 1 illustrates the two innovative system designs, engineered to facilitate systematic and equitable performance evaluation. Readers can refer to (Mallouh et al, 2022) for getting more details about configurations.

Configuration 1: Direct Water Circulation System.

Configuration 1, illustrated in Fig. 1(a), represents an active direct water circulation system characterized by direct water-solar collector interaction. The sophisticated system architecture encompasses critical components:

- Solar collector
- Fully insulated mixing chamber
- Precision proportional valve
- Variable-speed pump with dynamic flow rate modulation

The pump generates two distinct fluid streams: one routed through the proportional valve and another directed through the solar collector. This configuration is particularly well-suited for solar pool heating applications with extensive thermal storage capabilities.

Configuration 2: Thermal Energy Buffer System.

Configuration 2, depicted in Fig. 1(b), introduces a more complex thermal management approach. The system comprises:

- Solar collector
- Thermal storage tank
- Mixing chamber
- Two variable flow pumps
- Proportional valve

A distinctive feature of this configuration is its closed-loop design, where:

- One pump circulates water between the solar collector and storage tank
- A second pump transfers cold water from the input to the storage tank

The storage tank functions as a critical thermal energy buffer, providing substantial thermal capacity to stabilize outlet water temperature under variable solar radiation conditions. This design enables strategic thermal energy storage and controlled release, enhancing system resilience and performance.

Both configurations were subjected to rigorous temperature control protocols, ensuring precise and comparable performance assessment. The mathematical modeling, grounded in fundamental energy balance principles, provides a robust theoretical framework for: analyzing heat transfer mechanisms, evaluating thermal efficiency and characterizing system dynamics.

The system of Eq. 1 represents the mathematical model of configuration 1:

$$\begin{cases} C_c\dot{T}_c = A_cI_c\eta_c - A_cU_c(T_c - T_\infty) - c_w\dot{m}_c(T_{co} - T_{ci}) \\ C_{mc}\dot{T}_o = c_w\dot{m}_i(T_{co} - T_o) - c_w\dot{m}_v(T_{co} - T_i) \end{cases} \quad (1)$$

The system of Eq. 2 represents the mathematical model of configuration 2:

$$\begin{cases} C_c\dot{T}_c = A_cI_c\eta_c - A_cU_c(T_c - T_\infty) - c_w\dot{m}_c(T_{co} - T_{ci}) \\ C_s\dot{T}_s = c_w\dot{m}_c(T_{co} - T_{ci}) - c_w\dot{m}_s(T_s - T_i) - A_sU_s(T_s - T_\infty) \end{cases} \quad (2)$$

where:

T: Temperature [°C].

- Tc: Temperature of the solar collector [°C]
- Tci: Inlet temperature of the solar collector [°C]
- Tco: Outlet temperature of the solar collector [°C]
- To: Outlet temperature of the system (after the mixing chamber) [°C]
- Ti: Inlet temperature of the system [°C]
- T∞: Surrounding or ambient temperature [°C]
- Ts: Temperature of the storage tank [°C]

$\dot{m}$: Mass flow rate [kg/s]

- $\dot{m}_c$: :Mass flow rate through the solar collector [kg/s]
- $\dot{m}_s$: :Mass flow rate through the storage tank [kg/s]
- $\dot{m}_l$: Mass flow rate of the liquid stream [kg/s]
- $\dot{m}_v$: Mass flow rate of the vapor stream [kg/s].

U: Overall heat transfer coefficient [W/m^2·K].

- U_c: : Overall heat transfer coefficient of the solar collector [W/m^2·K]
- U_s: : Overall heat transfer coefficient of the storage tank [W/m^2·K]

η: Solar collector efficiency [%].
ηc: Efficiency of the solar collector [%]

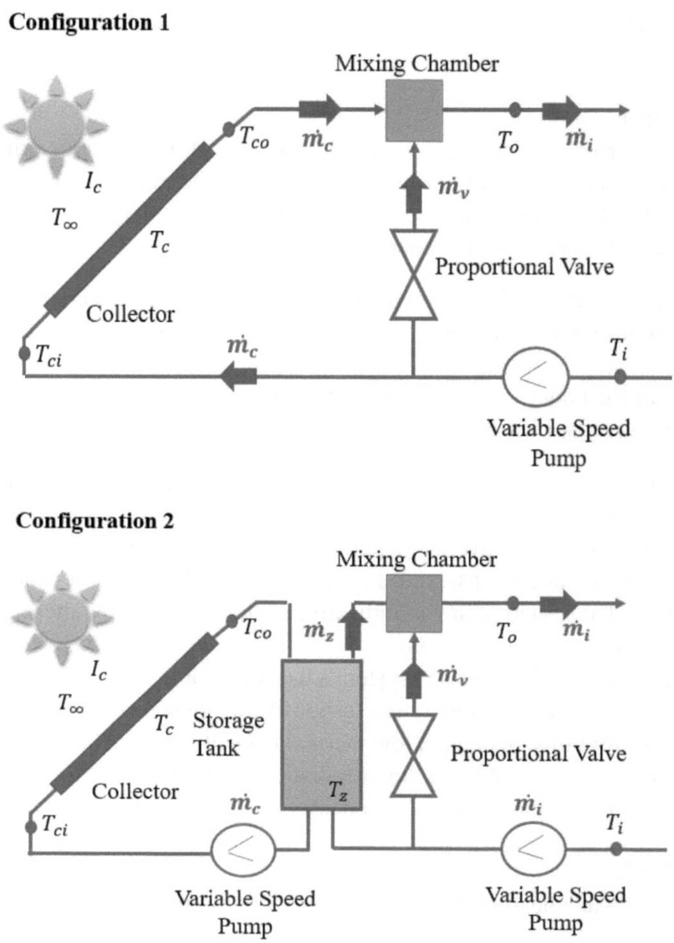

Fig. 1. Solar water heating systems with different configurations

C: Thermal mass capacity [kJ/K].

- C_c: Thermal mass capacity of the solar collector [kJ/K]
- C_s: Thermal mass capacity of the storage tank [kJ/K]
- C_{mc}: Thermal mass capacity of the mixing chamber [kJ/K]

c_w: Specific heat of water [kJ/kg·K]

3 Simulation Results

MATLAB Simulink is used for showing the temperatures variation in the case of two configuration.

The parameters are presented in Table 1and are selected according to (Mallouh et al. 2022).

Configuration 1

The system's thermal performance under clear sky conditions reveals critical insights into its operational dynamics (See Fig. 2). With predefined mass flow rates (inlet water: 0.275 kg/s, hot water: 0.1 kg/s, and steam: 0.175 kg/s), the outlet water temperature (To) demonstrated remarkable thermal characteristics:

- The temperature Range oscillated between 2 °C and 79 °C throughout the day.
- The maximum temperature approached 85.5 °C, near the system's thermal design limits.
- The thermal Responsiveness is characterized by a dramatic temperature fluctuation correlating directly with solar irradiation intensity.

Transitioning to cloudy atmospheric conditions dramatically transformed the system's thermal behavior as depicted in Fig. 3: The Outlet Water Temperature reached a maximum with 2 °C and an average equal to 61 °C. A significant reduction compared to clear sky performance is observed. Furthermore, the Tc shown a peak temperature reached 189 °C. The variation are an indicative of the system's low thermal inertia and high sensitivity to radiation variations.

From these numerical results, the system's thermal profile exhibits three fundamental characteristics: First, a rapid temperature responsiveness where an extreme sensitivity to solar radiation variations. Second, a low thermal capacity that enables quick adaptation to atmospheric changes. Finally, a direct radiation correlation that allows temperature profiles closely mirror solar radiation intensity.

Configuration 2

The second solar water heating (SWH) system was configured with precise mass flow rate

Table 1. Simulation parameters values utilized in the two SWH systems.

Parameter	Description	configuration 1	configuration 2
Ac[m^2]	Area of the solar collector	62.5	62.5
As[m^2]	Storage tank area	NA	4.7
C_w [kj/kg.K]	Specific heat capacity of water	4.2	4.2
C_c [kj/K]	Heat capacity of the solar collector	699.5	699.5
Cmc [kj/K]	Heat capacity of the mixing chamber	4.2	4.2
Cs[kj/K]	Heat capacity of the storage tank	NA	2.67
T_∞ [°C]	Ambient temperature	31	31
η_c	Efficiency of the solar collector	0.8	0.8
Uc[W/m^2.K]	Heat loss coefficient of the solar collector	7	7
Us[W/m^2.K]	Heat loss coefficient of the storage tank	NA	0.3

parameters where: The collector mass flow rate ($\dot{m}_c$): 3.5 kg/s and the Inlet water mass flow rate ($\dot{m}i$) is equal to 0.265 kg/s. These parameters were strategically established to investigate thermal performance under varying solar radiation conditions.

Under clear sky radiation, as illustrated in Fig. 4, the system demonstrated distinctive thermal characteristics: To is observed with an average temperature equal to 44 °C and a maximum temperature equal to 88 °C. We can speak about a thermal stability. Notably slower temperature transitions due to substantial thermal mass.

In the other hand, the cloudy atmospheric conditions significantly impacted system performance as presented by Fig. 5: The To can be observed with an average temperature equal to 35 °C and reaches a maximum temperature 69 °C.

The system's large thermal capacity fundamentally modifies its thermal response:

- Reduced temperature fluctuation rates.
- Enhanced thermal inertia.
- Slower adaptation to solar radiation variations.

In the other hand, a critical observation regarding temperature control. The desired outlet temperature maintained only briefly during clear sky conditions and a complete failure to achieve desired temperature under cloudy scenarios.

These observed thermal performance analysis reveals significant system design challenges:

- Complex relationship between thermal mass and temperature control.
- Limited adaptability to atmospheric variability.
- Need for advanced thermal management strategies.

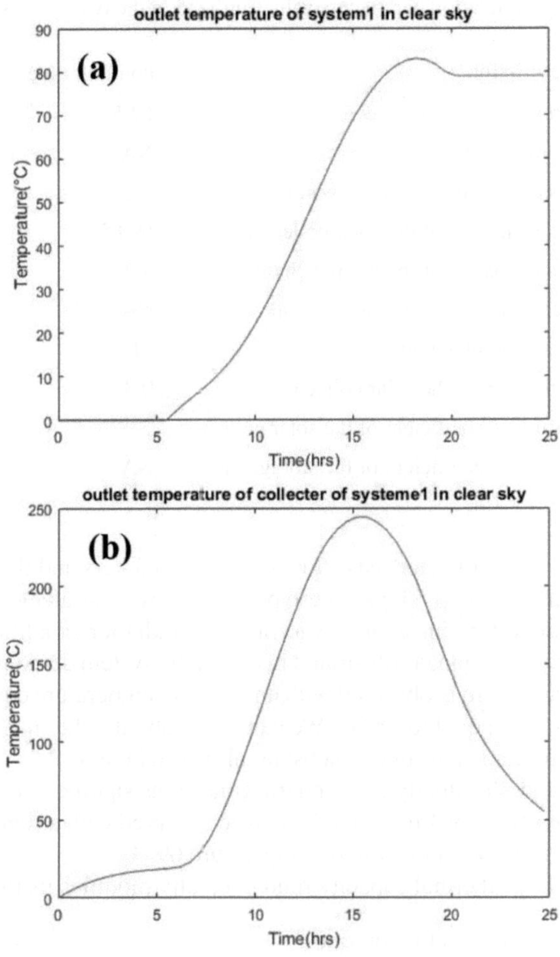

Fig. 2. Configuration 1 under a clear sky conditions (a) The outlet temperature of all the system, (b) The outlet temperature of collector

The performance analysis of two distinct solar water heating (SWH) system configurations reveals complex and nuanced insights into thermal energy management, highlighting critical intersections between solar radiation dynamics, system engineering, and thermal efficiency. The first system configuration demonstrated exceptional responsiveness to solar radiation, characterized by:

- Rapid temperature adaptation
- High sensitivity to environmental variations
- Potential for advanced thermal energy harvesting

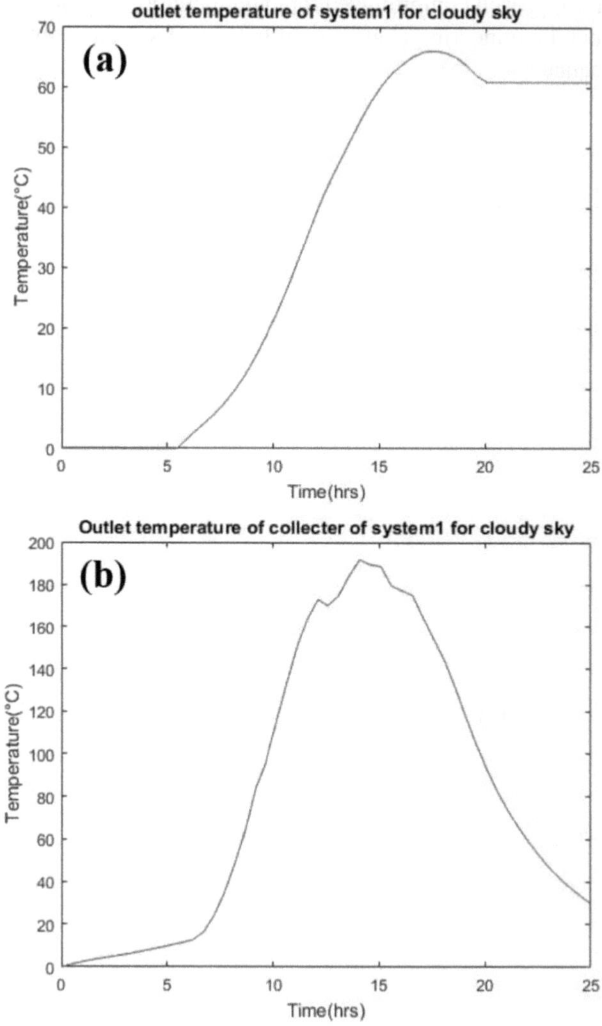

Fig. 3. Configuration 1 under cloudy sky conditions (a) The outlet temperature of all the system, (b) The outlet temperature of collector

While exhibiting remarkable dynamic capabilities, the configuration also exposed significant technological development opportunities in solar water heating system design.

The second system configuration emphasized thermal stability through:

- Substantial thermal inertia
- Enhanced energy storage capabilities
- Controlled temperature modulation

However, this approach compromised rapid temperature responsiveness, underscoring the fundamental engineering challenge of balancing system stability with adaptive thermal performance.

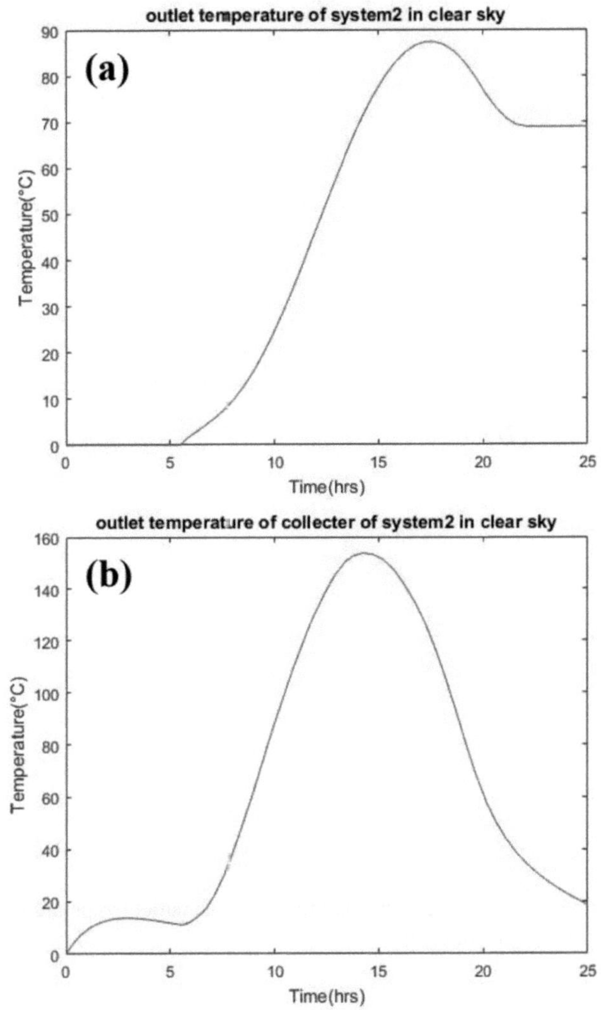

Fig. 4. Configuration 2 under a clear sky conditions (a) The outlet temperature of all the system, (b) The outlet temperature of collector

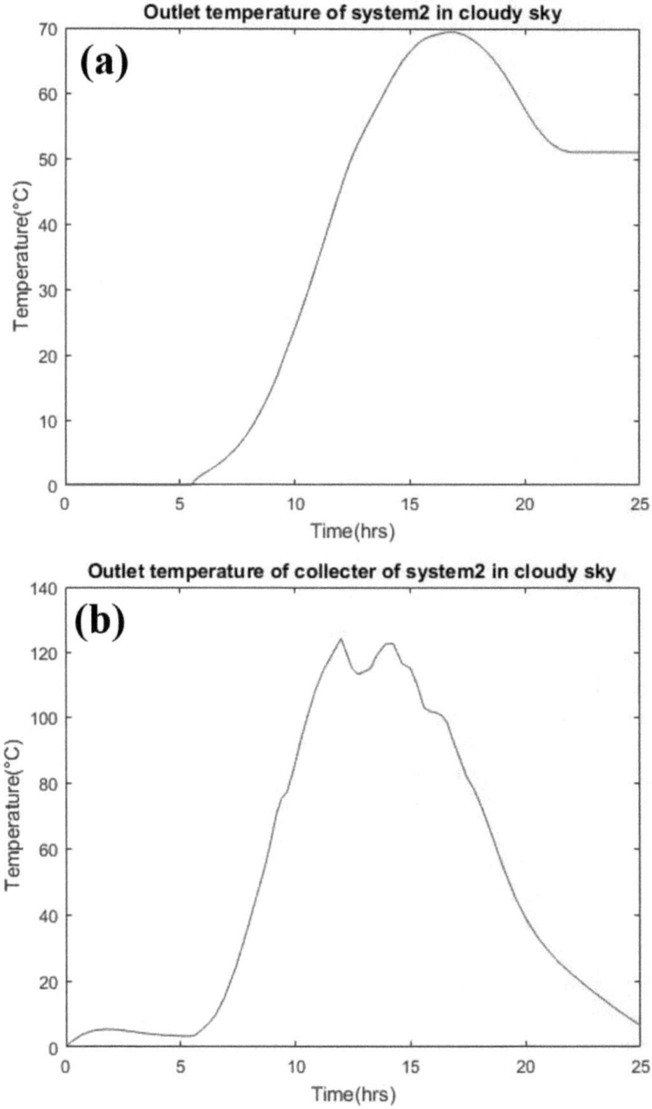

Fig. 5. Configuration 2 under cloudy sky conditions (a) The outlet temperature of all the system, (b) The outlet temperature of collector

4 Conclusion

The primary objective of solar water heating (SWH) systems is to maintain the outlet water temperature (To) at a desired 75 °C throughout the operational day. Comprehensive analysis of the two proposed system configurations under clear and cloudy sky conditions revealed significant performance limitations. Under clear sky conditions, was kept at the temperature that was desired for only a few minutes, while during cloudy conditions, the systems completely failed to regulate the temperature. Neither configuration successfully maintained the temperatures of critical components within acceptable operational limits, underscoring the complexity of thermal management in solar water heating technologies. Recognizing these limitations, the research will advance by implementing a closed-loop system simulation with two PID controllers to develop more precise temperature control strategies. While the initial results provide logical insights into system performance, they fundamentally highlight the critical need for advanced control mechanisms to achieve reliable and consistent thermal management in solar water heating systems.

Acknowledgments. The authors gratefully acknowledge the assistance and financial support provided by the project 3P6D2023-PEJC from the Tunisian Ministry of Higher Education and Scientific Research.

Refer The authors also acknowledge the support of the Tunisian-Mauritanian project n° "23MT1003" financed by the Tunisian Ministry of Higher Education and Scientific Research.

References

Odeh, S., Aden, I.: Modeling of a photovoltaic/thermal hybrid panel for residential hot water system. J. Solar Energy Eng. **147**(1) (2025)

Salah, S.I., Eltaweel, M., Abeykoon, C.: Towards a sustainable energy future for Egypt: a systematic review of renewable energy sources, technologies, challenges, and recommendations. Cleaner Eng. Technol **8**, 100497 (2022)

Meena, C.S., Prajapati, A.N., Kumar, A., Kumar, M.: Utilization of solar energy for water heating application to improve building energy efficiency: an experimental study. Buildings **12**(12), 2166 (2022). https://doi.org/10.3390/buildings12122166

Mallouh, M.A., AbdelMeguid, H., Salah, M.: A comprehensive comparison and control for different solar water heating system configurations. Engi. Sci. Technol. An Int. J. **35**, 101210 (2022). https://doi.org/10.1016/j.jestch.2022.101210

Rejab, M.N., Marwah, O.M.F., Johar, M.A., Ribuan, M.N.: Real-time thermal energy harvesting from solar radiation in Malaysia at low-temperature difference. J. Adv. Res. Fluid Mech. Thermal Sci. **107**(2), 117–132 (2023)

Ko, M.J.: Analysis and optimization design of a solar water heating system based on life cycle cost using a genetic algorithm. Energies **8**(10), 11380–11403 (2015). https://doi.org/10.3390/en81011380

Influence of Tank Top-Bottom Shape of a Solar Water Storage Tank on the Thermal Stratification

Ahmed Samet[1,2](✉)

[1] Laboratory of Mechanics, Modelling and Production, National School of Engineering of Sfax, University of Sfax, Sfax, Tunisia
samet.ahmed@yahoo.com
[2] Higher Institute of Applied Sciences and Technology of Gabes, University of Gabes, Gabes, Tunisia

Abstract. This work deals with the thermal stratification problem inside a vertical storage tank used in a Solar Water Heating System (SWHS). The study draws on established Unsteady 3-Dimensional Computational Fluid Dynamics (3D-CFD) tools, and results are validated with published experimental work. However, some changes on tank top-bottom shapes are made, and the stratification behavior is accordingly examined. Although these changes may seem a bit minimal, they are very interesting since they have a quite noticeable impact on the storage tank efficiency, and the most performing designs could be consequently selected. The 3D-CFD results of all proposed designs are compared through some mixture and stratification evaluation criteria, such as the outlet temperature. The general trend is that a spherical top-bottom shape produces a superior thermal performance.

Keywords: Computational Fluid Dynamics CFD · Stratification · Solar storage tank · Solar water heating system · Tank top-bottom shapes

1 Introduction

Improving the Solar Water Heating System (SWHS) capabilities continues to get special interest from both academia and industry. A Water Storage Tank (WST) is a prominent SWHS key equipment, which not only ensures thermal storage but also affects the overall system performance. Improving the thermal stratification and maintaining the stability of the thermocline are essential for enhancing the efficiency of the WST. In fact, the thermocline thickness determines the stratification quality. This thickness depends upon many parameters, namely the geometry of the WST, the water inlet, the fluid's thermo-hydraulic properties inside the WST and the configuration of obstacles within the tank. An ideal stratification is reached for a zero-thermocline thickness with uniform hot and cold temperature zones. Indeed, the storage performance is better as long as highest and lowest temperatures are sufficiently different (Haller et al. 2009, Haller et al. 2010, Castell et al. 2010, Hayat et al. 2017, Balotaki et al. 2019, Mawire et al. 2017, Prisecaru et al. 2017).

A. Hammami et al. (Eds.): MOSCOSSEE 2024, ACM 23, pp. 109–117, 2025.
https://doi.org/10.1007/978-3-032-00930-2_10

Several investigations have been carried out with the aim of improving the thermal stratification of the WST. Yaïci et al. (2013) studied the geometrical parameters effects of the WST, including variation in tank diameter, the positions of the inlet and outlet ports, the inlet variation temperature and the water temperature, finding that more the tank diameter decreases the more stratification is accomplished on the top tank layer. Other parameters effects have been examined by Bonanos et al. (2016) to improve the WST efficiency, namely the filter material properties and tank's height. Much of their work affirmed that the WST height makes a more significant effect on thermocline designs than fluid properties or particles size. Fernàndez Seara et al. (2007) studied the inlet and outlet ports arrangements effects on the thermal stratification. They gathered from their experiments that the inlet and outlet designs alter the WST performance. Shah et al. 2015 looked into the impact of the inlet stratifiers on the WST performance through numerical simulations and experiments. The proposed stratifier is a main tube with circular openings. A comparison between vertical tank and horizontal tank arrangements was provided by Abdelhak et al. (2015), and it was shown that, for a horizontal tank, the stratification degree drops in contrast to the vertical tank. A recent experimental investigation was performed by Assari et al. (2018) to evaluate the WST performance. Four test cases were considered, and indicated that the temperature of the tank bottom was inversely related to the collector efficiency. Kurşun et al. (2018) put finger on the thermal stratification behavior with regard to the tank orientation. They mentioned that the WST performance has been increased for a tank with an inclined placement. Furthermore, Assari et al. (2018) went into the details of the inlet-outlet locations effect on the WST performance through numerical and experimental studies. It was shown that the inlet hot water should be at the tank top, while the outlet cold water is rather directed towards the tank bottom.

It is worth mentioning that the state-of-the-art involved, solely, storage tanks with horizontal top-bottom shapes, and much of the aforementioned work has not yet explored a broader range of operating conditions This work will treat few added scenarios including new top-bottom shapes to improve the WST performance and pick the optimum design. The 3D- unsteady CFD are utilized to this end, and discussions of the obtained results with regard to the apported modifications are given accordingly.

2 Mathematical Formulation

The Navier-Stokes and energy equations formulations, are used for the temperature and velocity fields inside the WST (Samet et al. 2019):

The continuity equation:

$$\nabla.u = 0 \tag{1}$$

The momentum equation:

$$\rho\frac{\partial u}{\partial t} + (\rho u.\nabla)u = -\nabla p + \nabla.\tau - \rho g\beta\Delta T \tag{2}$$

The energy equation:

$$\rho C_p\frac{\partial T}{\partial t} + \rho C_p u.\nabla T = \nabla.(k\nabla T) \tag{3}$$

where $\tau = [\nabla u + \nabla u^T]$ stands for the stress tensor and $\rho g \beta \Delta T$ is the force of gravitational body, with ∇ denoting the gradient operator.

3 Computational Domains and CFD Validation

This section describes the computational domain and operating conditions outlined by Zachár et al. (2003). The Ansys Fluent CFD simulation process involves two main steps: first, constructing a WST geometry modeled after that presented by Zachár et al. (2003), and second, generating the mesh. Particular attention is given to the meshing process to ensure accurate simulation results. Various mesh sizes were tested, ranging from coarse to fine, until the numerical results became independent of mesh size. The numerical results were then compared with the experimental data reported in (Zachár et al. 2003), focusing on the temperature measurement line illustrated in Fig. 1.

A total number of cells equal to 430 000 and a 4 mm size of cell with a 10^{-3} s time increment are found to be appropriate and are chosen in the simulations. Concerning the solver details, the SIMPLE algorithm is used for flow computation and the problem solution is provided using implicit method. The pressure, momentum, energy, turbulent kinetic energy, and turbulent dissipation rate equations are discretized using the standard method and the first upwind method. In the current work, the geometry of the WST is symmetrical. Therefore, it is convenient to model only a half cylinder 3D CFD domain, as it will considerably reduce the computational time. Moreover, this will be suitable for simulating the cycles of charging and discharging which occur during water consumption. The following assumptions apply for the 3D-heat transfers and flow models, and are taken into account in the upcoming considered test cases:

- The state is unsteady
- The liquid is Newtonian
- The flow is supposed to be turbulent, in accordance with the calculated Reynolds number
- The gravitation was taken into account
- No slip condition is considered within the wall boundary
- The fluid properties remain the same except for fluid density variation with temperature in the buoyancy terms of the momentum equations which is treated using Boussinesq approximation.

Figure 2 tracks the fluid temperature variation along the WST height. It is shown that numerical results and experimental data match quite well, and remain within a 2% average temperature discrepancy. Three particular zones are clearly distinguished: A region characterized by a constant cold temperature referred to as cold-water zone. A second region corresponding to the thermocline region in which the stratification degree is important. Meanwhile, a third region is characterized by a hot water temperature.

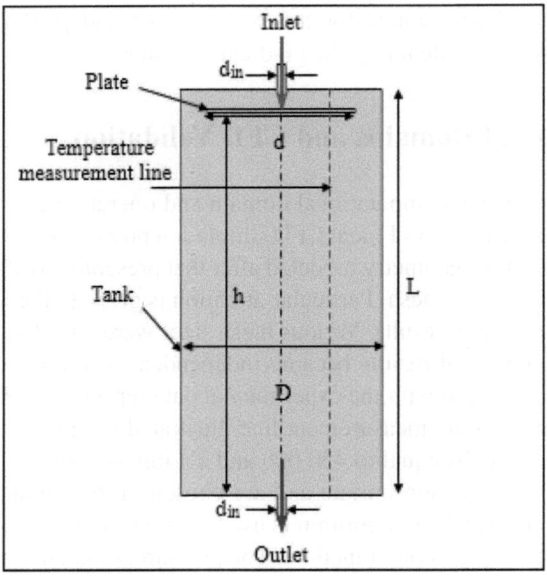

Fig. 1. The geometric configuration of the WST.

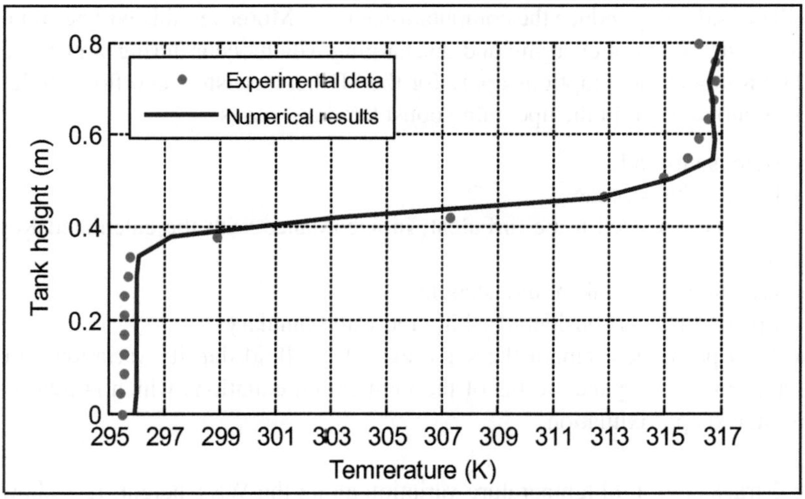

Fig. 2. Validation process: red dots for experimental data, blue full line for numerical results.

4 Results and Discussion

Hereafter, the impact of WST top-bottom shape on the thermal behavior is shown. Three cases are performed: the first WST with a horizontal shape as of Zachár et al. (2003), the second one with a conic shape and the third with a spherical shape in the top-bottom of WST as showed in Fig. 3.

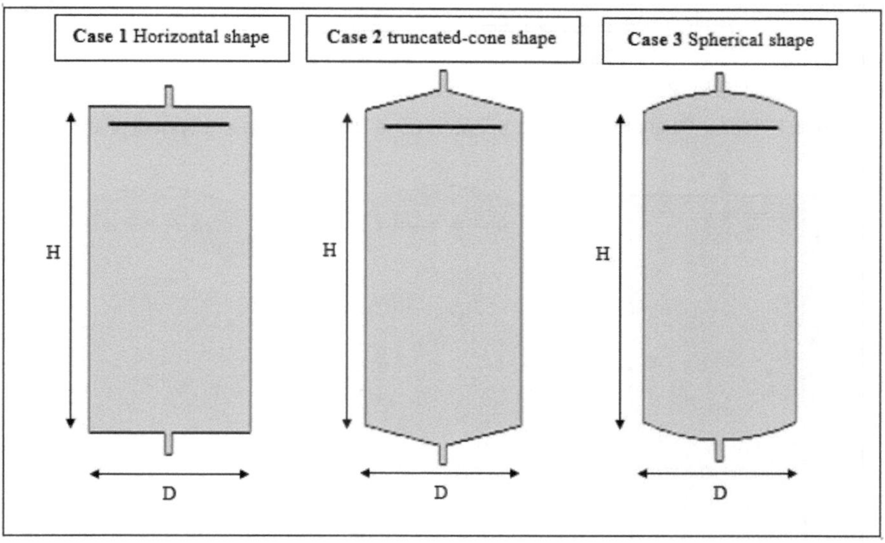

Fig. 3. Physical arrangements of the tank top-bottom shapes.

The objective of this section is to study the WST top-bottom shapes effect on the WST performance so as to obtain the tank optimum design.

To this intent, the effect of the tank top-bottom shapes on the repartition of temperature contours is highlighted. Figure 4 presents the temperature contours for the different storage tank. The hot water diffused with high momentum inside the WST causes a mixture flows over the upper layers. This is confirmed by Fig. 4 (a) at t = 700s.

As time goes by, three distinct thermal regions develop within the WST. A hot water region forms at the top of the WST, while a cold-water region persists near the bottom. A thermocline divides these two regions which tends to move along the WST height from the top towards the bottom. This phenomenon is obvious in all cases at t = 1800s, as seen in Fig. 4 (b).

The impact of the WST top-bottom shapes on the WST outlet temperature is shown in Fig. 5. The analysis indicates that the optimal design performance is achieved for the WST with spherical top-bottom shape. However, the use of horizontal top-bottom shape reduced the ability to sustain a distinct thermocline. On the other hand, the use of spherical or truncated-cone shapes improves the outlet temperature evolution and improves the ability to maintain a sharp thermocline.

To confirm our claims, a comparison through calculation of the full efficiency was furthermore performed. The full efficiency (FE) is another important parameter used to qualify the thermal stratification (Wang et al. 2017).

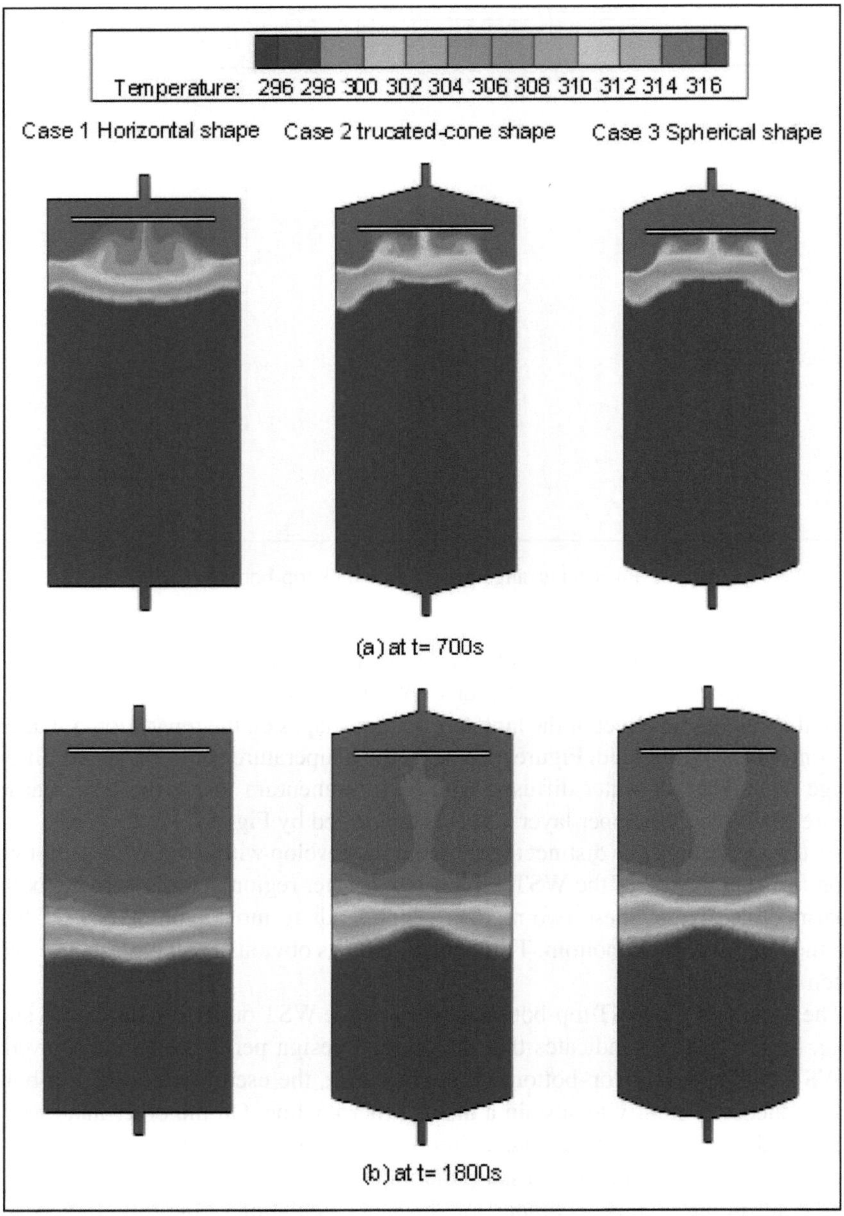

Fig. 4. Influence of tank top-bottom shapes on the temperature behavior.

Values corresponding to each of the three shapes are shown in Fig. 6. Highest value was obtained for the tank with spherical top-bottom shape (case 3), while a slightly less efficiency was porvided by the truncated-cone shape (case 2). Though, both design changes guaranteed a significant improvement to the horizontal top-bottom shape design (case 1) proposed by Zachár et al. (2003).

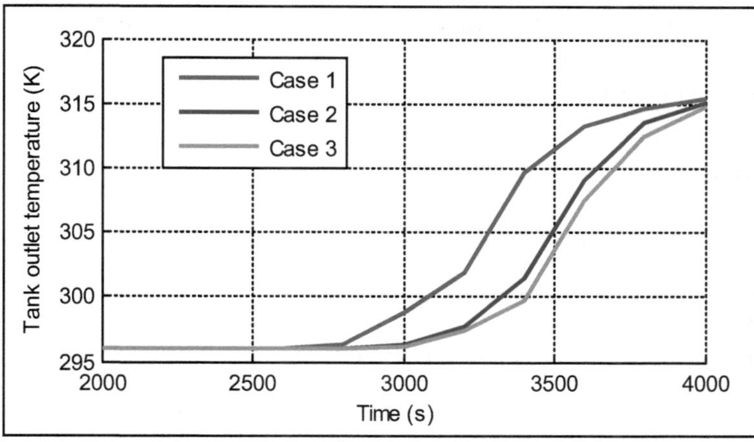

Fig. 5. Influence of tank top-bottom shapes on the outlet temperature. (Case 1: Horizontal shape, Case 2: truncated cone shape, Case 3: spherical shape).

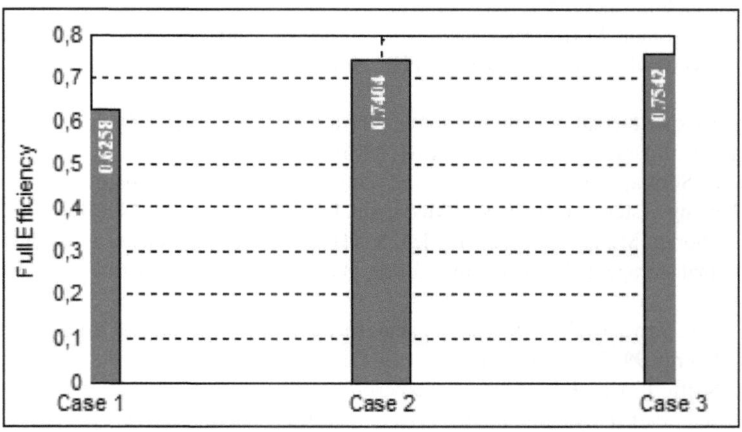

Fig. 6. Influence of tank top-bottom shapes on the Full efficiency. (Case 1: Horizontal shape, Case 2: truncated cone shape, Case 3: spherical shape).

It is already safe to say that tank top-bottom shapes had an influence on the efficiency of the storage tank, and best performance was achieved for the spherical shape. From now on, we will rather pick the optimum design of the tank (spherical shape) to study the influence of the plate shapes of the tank on the storage tank performance.

5 Conclusion

The aim of this work was to highlight the impact of tank top-bottom shape on the WST performance while being used for SWHS. The followed computational fluid dynamic code was briefly revisited. Throughout the work, different new configurations of tank top-bottom shapes were studied and compared against existing models. The comparisons clearly showed that the performance does depend on the WST designs and that the set expectations were justified. Indeed, a superior thermal performance was achieved for the WST with the spherical top-bottom shape.

References

Haller, M.Y., Yazdanshenas, E., Andersen, E., Bales, C., Streicher, W., Furbo, S.: A method to determine stratification efficiency of thermal energy storage processes independently from storage heat losses. Sol. Energy **84**, 997–1007 (2010)

Castell, A., Medrano, M., Solé, C., Cabeza, L.F.: Dimensionless numbers used to characterize stratification in water tanks for discharging at low flow rates. Renew. Energy **35**, 2192–2199 (2010)

Hayat, T., Mumtaz, M., Shafiq, A., Alsaedi, A.: Thermal stratified three- dimensional flow with inclined magnetic field and Joule heating. J. Braz. Soc. Mech. Sci. Eng. **39**(5), 1607–1621 (2017)

Haller, M.Y., Cruickshank, C.A., Streicher, W., Harrison, S.J., Andersen, E., Furbo, S.: Methods to determine stratification efficiency of thermal energy storage processes–review and theoretical comparison. Sol. Energy **83**, 1847–1860 (2009)

Balotaki, H.K., Saidi, M.H.: Design and performance of a novel hybrid photovoltaic–thermal collector with pulsating heat pipe (PVTPHP). Iranian J. Sci. Technol. Trans. Mech. Eng. 1–11 (2019)

Mawire, A.: Experimental de-stratification and heat loss in a storage tank con-taining different thermal oils. J. Braz. Soc. Mech. Sci. Eng. **39**(6), 2279–2288 (2017)

Prisecaru, T., Soriga, I.: Control strategy and thermal inertia influence on evacuated tube solar water heating systems. J. Braz. Soc. Mech. Sci. Eng. **39**(6), 2131–2144 (2017)

Yaïci, W., Ghorab, M., Entchev, E., Hayden, S.: Three-dimensional unsteady CFD simulations of a thermal storage tank performance for optimum design. Appl. Thermal Eng. **60**, 152–163 (2013)

Bonanos, A.M., Votyakov, E.V.: Sensitivity analysis for thermocline thermal storange tank design. Renew. Energy **99**, 764–771 (2016)

Fernandez-Seara, J., Uhı, F.J., Sieres, J.: Experimental analysis of a domestic electric hot water storage tank. Part II: dynamic mode of operation. Appl. Therm. Eng. **27**, 137–144 (2007)

Shah, L.J., Andersen, E., Furbo, S.: Theoretical and experimental investigations of inlet stratifiers for solar storage tanks. Appl. Therm. Eng. **25**, 2086–2099 (2005)

Abdelhak, O., Mhiri, H., Bournot, P.: CFD analysis of thermal stratification in domestic hot water storage tank during dynamic mod. Build. Simul. **8**(4), 421–429 (2015). https://doi.org/10.1007/s12273-015-0216-9

Assari, M.R., Tabrizi, H.B., Movahedi, M.J.: Experimental study on destruction of thermal stratification tank in solar collector performance. J. Energy Storage **15**, 124–132 (2018)

Kurşun, B., Ökten, K.: Effect of rectangular hot water tank position and aspect ratio on thermal stratification enhancement. Renew. Energy **116**, 639–646 (2018)

Assari, M.R., Tabrizi, H.B., Savadkohy, M.: Numerical and experimental study of inlet-outlet locations effect in horizontal storage tank of solar water heater. Sustain. Energy Technol. Assess. **25**, 181–190 (2018)

Samet, A., Ben Souf, M.A., Fakhfakh, T., Haddar, M.: Numerical investigation of the baffle plates effect on the solar water storage tank efficiency. Energy Sour. Part A: Recovery Utiliz.and Environ. Effects (2019)

Zachár, A., Farkas, I., Szlivka, F.: Numerical analyses of the impact of plates for thermal stratification inside a storage tank with upper and lower inlet flows. Sol. Energy **74**, 287–302 (2003)

Wang, Z., Zhang, H., Dou, B., Huang, H., Wu, W., Wang, Z.: Experimental and numerical research of thermal stratification with a novel inlet in a dynamic hot water storage tank. Renew. Energy **111**, 353–371 (2017)

The Use of Discrete Random Separation DRS for Fault Detection in a Wind Turbine Gearbox

Ayoub Mbarek(✉), Ahmed Hammami, and Fakher Chaari

Laboratory of Mechanics, Modeling and Production (LA2MP), National School of Engineers of Sfax, BP1173, 3038 Sfax, Tunisia
ayoubmbarekenit@gmail.com

Abstract. Gear elements are presented in a huge industrial and engineering application as well as in multiscale form like the actuator gear and the wind turbine transmission. They can work under different severe conditions. The gear transmissions are subjected to various kinds of excitations which can cause failure and mal function that can possibly lead to catastrophic scenario. In this work, a signal processing tool for damage detection of Wind turbine gearbox is investigated. In fact, the vibration responses generated by the developed numerical of the gearbox transmission are proceeded and treated using Discrete Random Separation (DRS) method. The obtained results show the ability of the proposed method to detect failure localized on the sun gear component. The method is simple to apply and does not require several parameters. It also shows a good capability to remove harmonics from the vibration response signal.

Keywords: wind turbine gearbox · defect · DRS · vibration signal

1 Introduction

A very important tool for analysing the condition of a machine is vibration measurement and signal processing [1–3].

In industrial conditions, the measurement process is very complex especially when it depicts highly complex machines or structures. The measured signal is generally rich with noise contaminated by noise, and it becomes hard and is very complicated to analyse. Thus, several signal processing algorithms have been developed. One of the methods allowing the separation of deterministic parts from the noisy components is the Time Synchronous Averaging (TSA) [4]. This method has been regenerated and enhanced later by several research papers like in the works of McFadden [5, 6]. Bonnardot et al. [7], and Coats et al. [8]. In addition, Antoni and Randall [9] developed a method based on the modification of kurtosis and spectral kurtosis. The technique provides an interesting demodulation center frequency which is in general rich with impulsive signals to facilitate the introduction of the envelope analysis. In [10], authors used Discrete Random Separation (DRS) method to cancel and remove noise from the signal.

In this work, the DRS was used and adapted to a wind turbine gearbox. Section 2 describes the theoretical background of this method, and Sect. 3 illustrates some results.

A. Hammami et al. (Eds.): MOSCOSSEE 2024, ACM 23, pp. 118–125, 2025.
https://doi.org/10.1007/978-3-032-00930-2_11

2 Discrete Random Separation DRS Method

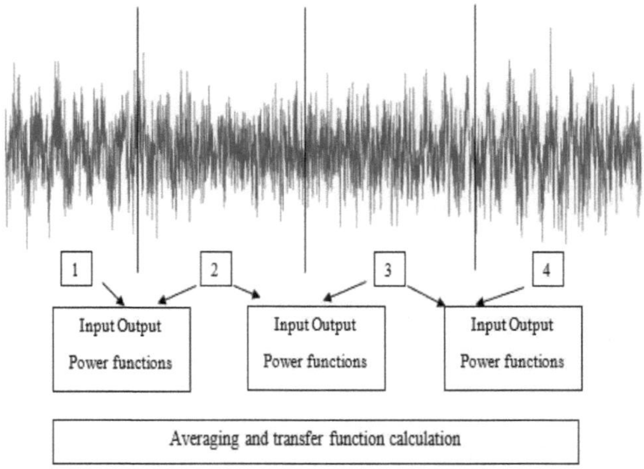

Fig. 1. Structure and illustration of the DRS method

The DRS technique is adopted to separate the deterministic parts and the random components of the measured signal. The rotating machinery could be excited or driven in stationary condition (fixed external speed and load). As mentioned in the introduction, this technique is well appropriate to rotating machinery systems. This process is detailed in Fig. 1, the transfer function can be determined using the following equation.

$$H_1(f) = \frac{G_{xy}(f)}{G_{xx}(f)} \tag{1}$$

where $G_{xy}(f)$ and $G_{xx}(f)$ represent respectively the cross-spectrum the auto-spectrum

$$G_{XY}(f) = \Im\{Y(t)\}\Im^*\{X(t)\} \tag{2}$$

$$G_{XX}(f) = \Im\{X(t)\}\Im^*\{X(t)\} \tag{3}$$

$\Im$ is the Fourier transform whereas *X(t)* and *Y(t)* are respectively the input and the output signals, and * presents the complex conjugate.

$$G_{XY}(f) = \frac{1}{N}\sum\nolimits_{i=1}^{N} \Im\{Y_i(t)\}\Im^*\{X_i(t)\} \tag{4}$$

$$G_{XX}(f) = \frac{1}{N}\sum\nolimits_{i=1}^{N} \Im\{X_i(t)\}\Im^*\{X_i(t)\} \tag{5}$$

The transfer function is then determined in the time domain by using the inverse Fourier transform:

$$h(t) = \Im^{-1}\{H_1(f)\} \tag{6}$$

With the application of this procedure, the measured original signal can be filtered.

In order to illustrate the DRS method, the same 40 Hz signal of unity amplitude, added by a zero mean Gaussian white noise of variance equal to 0.2 was used. The signal to noise ratio (SNR) is 3.98. The used signal is characteristic by a sampling frequency equal to 400 Hz and a total time record equal to 100 s. Figure 2 illustrate the used signal for the test.

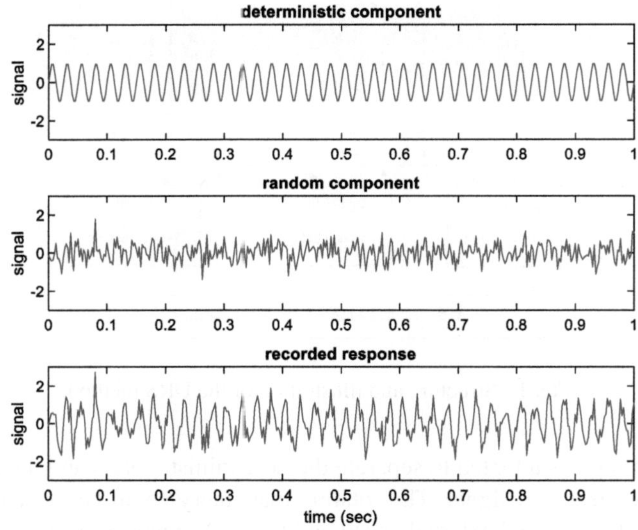

Fig. 2. Sine signal of 40 Hz, Gaussian white noise and mixed signal.

Moreover, Figs. 3 and 4 present the transfer function obtained with the proposed method.

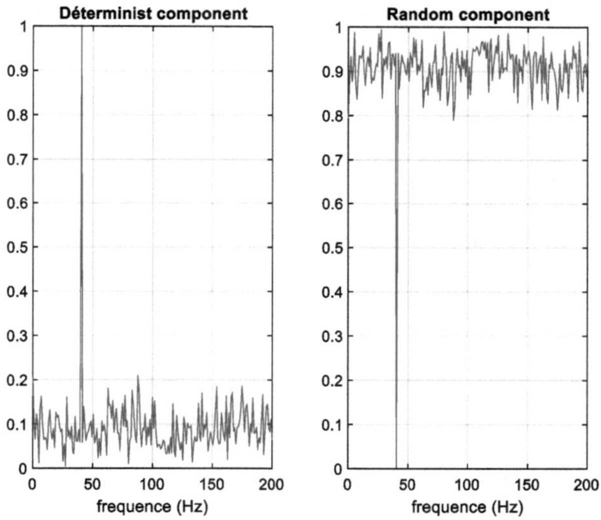

Fig. 3. Transfer functions obtained by the DRS method.

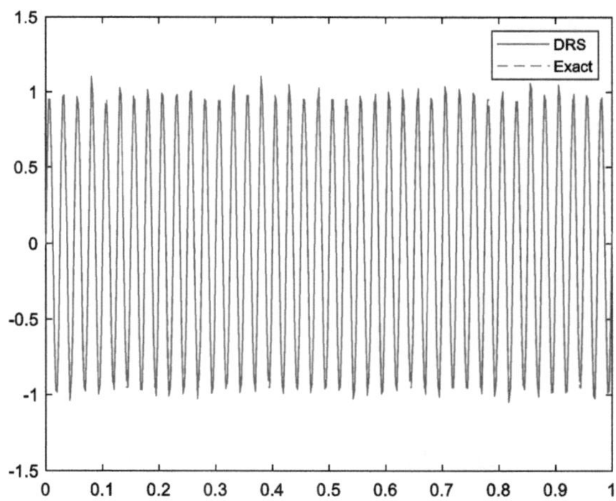

Fig. 4. The obtained determinist component by the DRS method versus exact signal.

3 Results

The DRS method was applied to vibratory accelerations signals simulated on a planetary gear system model. The numerical model was used in many research papers of the authors, it is well detailed in such as references like [11–15].

From this model, the accelerations signal of ring components was extracted, and its spectrum was determined as illustrated in Figs. 5 and 6.

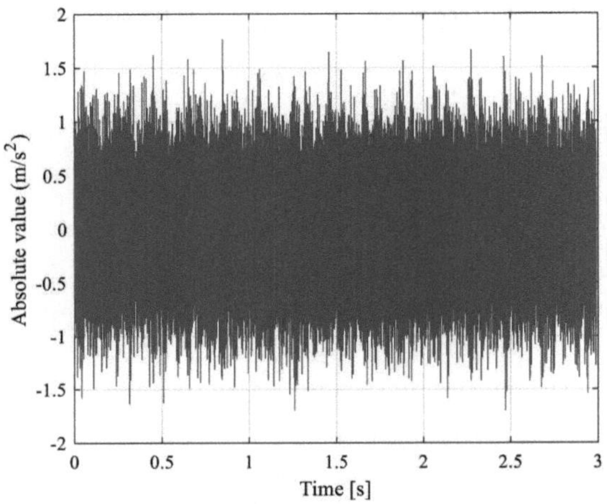

Fig. 5. The vibration response simulated on the y direction of the defected ring

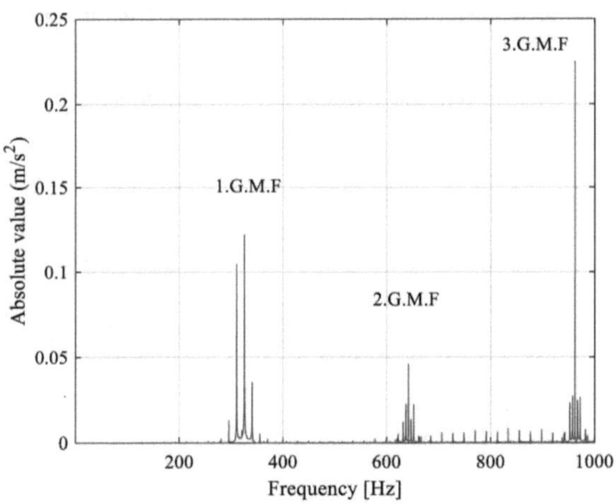

Fig. 6. Spectrum of the defected ring

Figure 7 and Fig. 8 present a zoom around the first and second gear mesh frequency **GMF** respectively. It can be noted that there is an important amplitude of peaks in damaged cases in comparison with the healthy ones. In Fig. 7, regular peaks in the health case are spaced of 5 Hz equal to ***fc***. However, only three peaks appear in damaged case, distanced each other of 20 Hz.

In the neighbour of the second GMF, the healthy responses do not give pronounced peaks. It is totally different with the damaged responses. Heavy peaks are highlighted in both sides of ***GMF*** regularly spaced of 5 Hz that is equal to ***fc***. In addition, two extra peaks distanced from ***GMF*** 20 Hz, indicate the presence of the sun cracked tooth as the frequency fds is equal to 20 Hz.

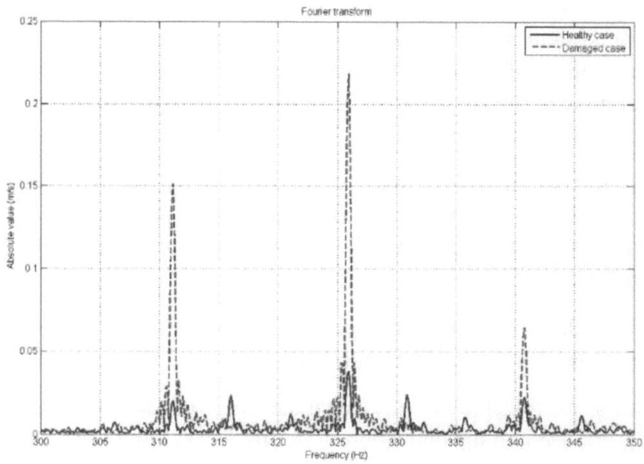

Fig. 7. [DRS method] Zoom around the first GMF

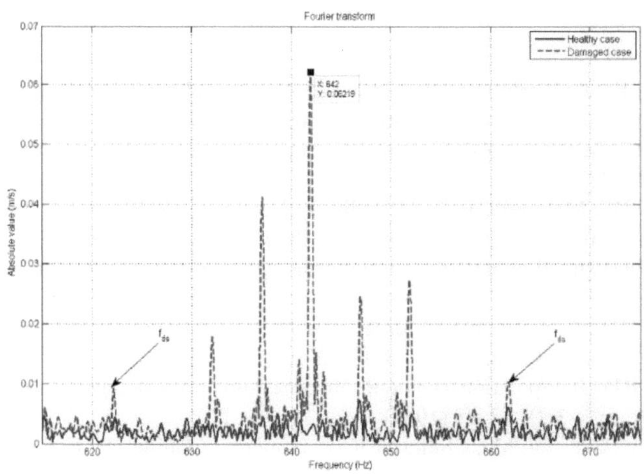

Fig. 8. [DRS method] Zoom around the second GMF

The DRS technique shows a significant result. The frequency of the faulty sun is well observed as well as the frequencies characteristics. The DRS method amplify the deterministic components and kept some of random waves.

4 Conclusion

Planetary gear fault detection problem was solved using powerful signal processing tools. The damaged tooth causes high vibrations amplitude. In the case of complex systems, which contain several components and different influent functions, the detection protocol is very complex and sensitive.

DRS signal processing tool was applied to the vibration response of a planetary gear system. The system was excited in healthy and in damaged tooth conditions. Both methods show a significant result for damage detection. In fact, the obtained deterministic components define well the frequencies features related to the structural dynamics.

Some conclusions can be retired:

1. The cracked defects affect the time recorded signal by the increasing of the signal amplitude in the dime and by the apparition of several peaks in the signal spectrum.
2. The application of DRS positively influenced the spectrum by amplifying deterministic components.

As future work, the numerical model and the obtained results will be used for machine learning objective and for prognostics purposes.

Tribute. Authors would like to pay attribute to Professor Fakher CHAARI and appreciate his guidance and his contribution in this work. Pay to his soul.

Acknowledgement. This work is financed by the Tunisian-Mauritanian project n° "23MT1003".

References

1. Dziedziech, K., Jablonski, A., Dworakowski, Z.: A novel method for speed recovery from vibration signal under highly non-stationary conditions. Measurement **128**, 13–22 (2018)
2. Ugweje, O.C.: Selective noise filtration of image signals using wavelet transform. Measurement **36**, 279–287 (2004)
3. Makowski, P.L., Morawski, R.Z., Michalik, Ł, Domanski, A.W.: An efficient algorithm for processing data from a multi-point sensor of vibrations. Measurement **44**, 2060–2067 (2011)
4. Braun, S.: The extraction of periodic waveforms by time domain averaging. Acoustica **23**(2), 69–77 (1975)
5. McFadden, P.D.: A revised model for the extraction of periodic waveforms by time domain averaging. Mech. Syst. Sig. Process. **1**(1), 83–95 (1987)
6. McFadden, P.D.: Interpolation techniques for time domain averaging of gear vibration. Mech. Syst. Sig. Process. **3**(1), 87–97 (1989)
7. Bonnardot, F., Badaoui, M.E., Randall, R.B., Danière, J., Guillet, F.: Use of the acceleration signal of a gearbox in order to perform angular resampling (with limited speed fluctuation). Mech. Syst. Sig. Process. **19**, 766–785 (2005)
8. Coats, M.D., Sawalhi, N., Randall, R.B.: Extraction of tacho information from a vibration signal for improved synchronous averaging. In: Proceedings of Acoustics Australia 2009, Adelaide, Australia, 23–25 November 2009
9. Antoni, J., Randall, R.: The spectral kurtosis: application to the vibratory surveillance and diagnostics of rotating machines. Mech. Syst. Signal Process. **20**(2), 308–331 (2006)
10. Antoni, J., Randall, R.B.: Unsupervised noise cancellation for vibration signals: part II – a novel frequency-domain algorithm. Mech. Syst. Sig. Process. **18**, 103–117 (2004)
11. Mbarek, A., Hammami, A., Del Rincon, A.F., Chaari, F., Rueda, F.V., Haddar, M.: Effect of load and meshing stiffness variation on modal properties of planetary gear. Appl. Acoust. (2017)
12. Mbarek, A., Fernández, A., Hammami, A., Chaari, F., Viadero, F., Haddar, M.: Operational modal analysis using order-based analysis in a two-stage planetary gear with mechanical power recirculation. Comptes Rendus. Mécanique **350**(G2), 391–411 (2022)

13. Mbarek, A., Hammami, A., Del Rincon, A.F., Chaari, F., Rueda, F.V., Haddar, M.: Mesh force analysis of planetary gear system with defect. In: Walha, L., et al. (eds.) CMSM 2021. LNME, pp. 680–686. Springer, Cham (2023). https://doi.org/10.1007/978-3-031-14615-2_76
14. Mbarek, A., et al.: Influence of the non-linear Hertzian stiffness on the dynamic behavior of planetary gear during run up condition. In: Fernandez Del Rincon, A., Viadero Rueda, F., Chaari, F., Zimroz, R., Haddar, M. (eds.) CMMNO 2018, vol. 15, pp. 30–38. Springer, Cham (2019). https://doi.org/10.1007/978-3-030-11220-2_4
15. Mbarek, A., Fernández Del Rincon, A., Hammami, A., Chaari, F., Viadero, F., Haddar, M.: Early damage detection in planetary gear transmission in different operating conditions. Forsch. Ingenieurwes. **86**(4), 861–874 (2022)

Contribution to the Study of Dynamic Behavior of Wind Turbine Gearbox

Ahmed Hammami[1,2(✉)], Ayoub Mbarek[1], Fakher Chaari[1], Abdel Kader Mahmoud[3], Mohamed Haddar[1], and Ahmed Mohamed Yahya[3]

[1] Laboratory of Mechanics, Modeling and Production (LA2MP), Sfax, Tunisia
ahmed.hammami2109@gmail.com
[2] National School of Engineers of Gabes, 6029 Gabes, Tunisia
[3] Research Unit URA3E - EX CRAER, Nouakchott, Mauritania

Abstract. Wind turbines are a promising source of renewable energy. Due to the energy-intensive nature of our daily activities, reducing operating and maintenance costs is priorities in wind turbine operation. In fact, Wind Turbine Gearboxes (WTG) often operate under variable load and speed conditions, where there are problems with component frequency identification and fault diagnosis. The objective of this paper is to study the aerodynamic effects on the dynamic behaviour of WTG with the aim of providing a useful tool for designers and maintenance teams in the wind turbine industry. A dynamic model of a WTG is proposed. This gearbox consists of a planetary gear connected to the blade by an elastic coupling and a two-stage spur gear driving the generator. The aerodynamic torque is induced as an external force in the equations of motion, which are calculated using the Newmark algorithm. In addition, the stiffness of the gear mesh, which models the contact between the teeth in all wind turbine gearboxes, is sensitive to the speed variation of the rotor, which depends on the wind behaviour. The numerical results are shown in the time-domain and in the time-frequency domain as wind turbines operate in non-stationary conditions.

Keywords: WTG · Aerodynamic torque · Dynamic response · Gear mesh stiffness

1 Introduction

Vibration analysis of gearbox is very common approach for condition monitoring of rotating machinery (Chakroun et al., 2021), especially wind turbine gearboxes (Zimroz et al., 2011). It requires advanced signal processing for the case of wind turbine transmissions operating under transient conditions (Ziani et al., 2019).

In fact, speed and torque variations change the frequency of the gearbox vibration signals. Most of the works focused on helical gearboxes. Abboudi et al. (Abboudi et al., 2011) studied the dynamic behaviour of a WTG consisting of a two-stage spur gearbox. Then, Hadj Kacem et al. (Hadj Kacem et al., 2021) considered the uncertainty of some parameters in studying the dynamic behaviour of the same WTG. Zghal et al. (Zghal et al., 2019) investigated the dynamic behaviour of a bevel gear wind turbine gearbox

A. Hammami et al. (Eds.): MOSCOSSEE 2024, ACM 23, pp. 126–134, 2025.
https://doi.org/10.1007/978-3-032-00930-2_12

in the transient regime. Their proposed dynamic model is based on the work of (Driss et al., 2014).

Planetary gearboxes are increasingly used to drive large wind turbines. For the work done on the planetary wind turbine gearbox, we can cite the work of Viadero et al. (Viadero et al., 2014) proposed a multi-body model to study the dynamic behaviour of a wind turbine power train during start-up and emergency stop. Dewangan et al. (Dewangan et al., 2020) carried out the modal analysis of a WTG consisting of a two-stage planetary gearbox and a single-stage spur gearbox, and studied its dynamic behaviour under stationary conditions to identify the frequency components (Dewangan et al., 2022a) and under variable speed (Dewangan et al., 2022b). These works are based on research focused on the dynamic behaviour of planetary gears running in non-stationary conditions. Chaari et al. (Chaari et al., 2013) studied the effect of variable load on the dynamic response of planetary gearboxes where the input speed is sensitive to variable load. Hammami et al. correlated the numerical vibration signal with the experimental one of a back-to-back planetary gearbox during run-up (Hammami et al., 2021) and run-down (Hammami et al., 2015) regimes and in the presence of defects (Mbarek et al., 2022).

In this work, the estimated vibration signal of another WTG configuration is processed. It consists of the driven rotor, an elastic coupling, a planetary gear and a two-stage spur gear. The aerodynamic effect is taken into account in this study.

2 The Dynamic Model of the Mechanical Transmission of the Wind Turbine

The WTG consists of the driven rotor, an elastic coupling, a planetary gear and a two-stage spur gear. In the dynamic model, three d-o-f components are considered and the aerodynamic effects are taken into account (Fig. 1).

The Kramer model (Kramer, 1993) is adopted for the elastic coupling as it predicts its dynamic response well (Tadeo et al., 2011; Hmida et al., 2019). In the planetary gear, the sun-planet and planet-ring contact is modelled by the gear mesh stiffness K_{SMP} and K_{MPR} respectively. The two-stage helical gearbox consists of four gears. Gears (11) and (12) are connected by mesh stiffness K_{M1}, gears (21) and (22) are connected by mesh stiffness K_{M2}. Five shafts, introduced into the model through their torsional stiffness $K_{\theta i}$ ($i = 1..5$), connect the inertias of each block, which is supported by a bearing modelled by two springs K_{Xi} and K_{Yi} ($i = 1..5$).

The d-o-f vector is defined:

$$q = \{q_1, q_2, q_3\} \tag{1}$$

where:

$$q_1 = \{X_1, Y_1, \theta_W, \theta_{C1}, X_2, Y_2, \theta_{C2}, \theta_{CP}\} \tag{2}$$

$$q_2 = \{X_{P1}, Y_{P1}, \theta_{P1}, X_{P2}, Y_{P2}, \theta_{P2}, X_{P3}, Y_{P3}, \theta_{P3}\} \tag{3}$$

$$q_3 = \{X_3, Y_3, \theta_S, \theta_{11}, X_4, Y_4, \theta_{12}, \theta_{21}, X_5, Y_5, \theta_{21}, \theta_G\} \tag{4}$$

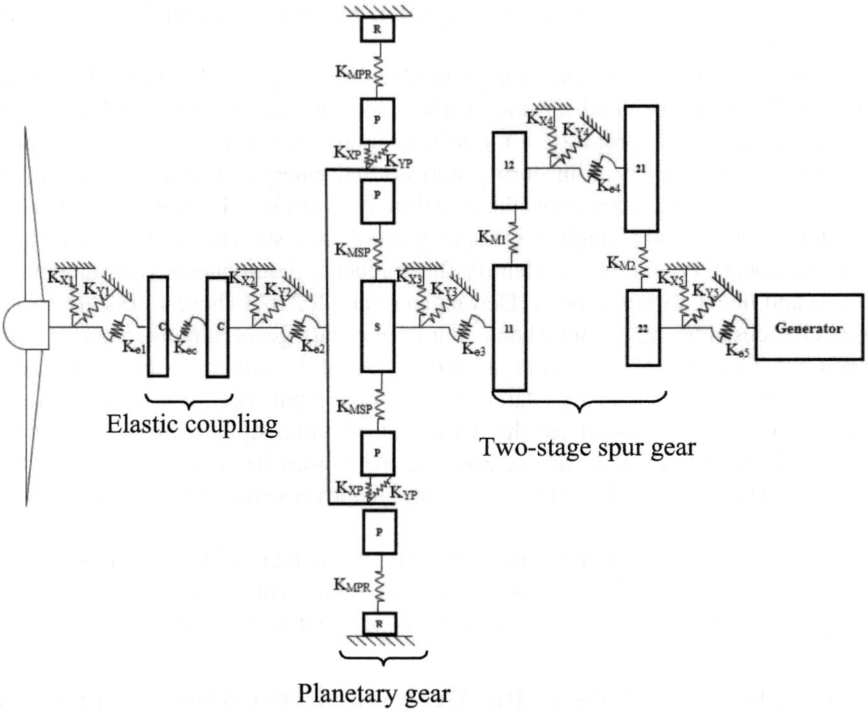

Fig. 1. Dynamic model of WTG

Table 1. Parameters of the dynamic model

Planetary gear	Carrier	Ring	Sun	Planet
Teeth number	-	66	18	24
Module (mm)	-	4		
Mass (Kg)	3.6	28.1	0.485	1.225
Inertia (10–3 kg.m^2)	21.5	697.8	0.35	2.04
Spur gear	**Wheel**		**Pinion**	
Teeth number	20		40	
Module (mm)	4			
Inertia (10–3 kg.m^2)	1.2		3.2	
Motor, receiver and coupling	**Motor**	**Receiver**	**Elastic coupling**	
Inertia (10–3 kg.m^2)	5.6	6	4	

The parameters of the dynamic model are summarized in Table 1.

Lagrange formalism is applied to this model in order to obtain the equations of motion:

$$M\ddot{q} + C\dot{q} + (K_s + K(t))q = F(t) \tag{5}$$

M is the mass matrix, K_s is the structure matrix. $K(t)$ is the time varying matrix. C is the proportional damping matrix.

3 Modelling of Aerodynamic Effects

The wind speed oscillates randomly around 7 m/s (Fig. 2).

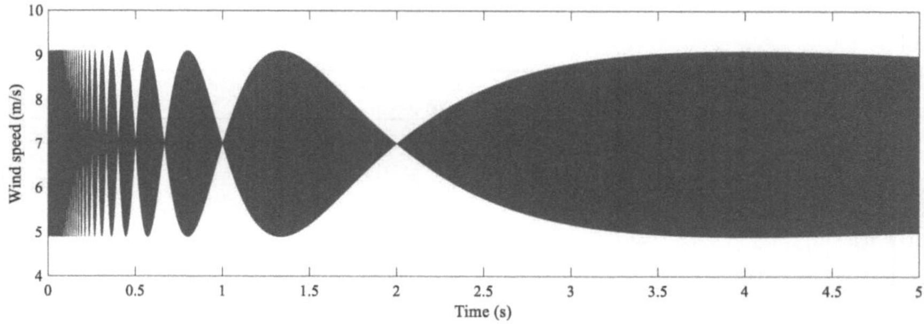

Fig. 2. Evolution of wind speed

The wind torque is defined as follows:

$$Q = \frac{N.R}{2\pi} \int_0^{2\pi} C_T.\rho.C.H.w^2 d\theta \tag{6}$$

N, R, C and *H* are respectively the number, the radius, the chord and the height of blades. ρ is the air density and w is the wind speed. C_T is a tangentiel coefficient.

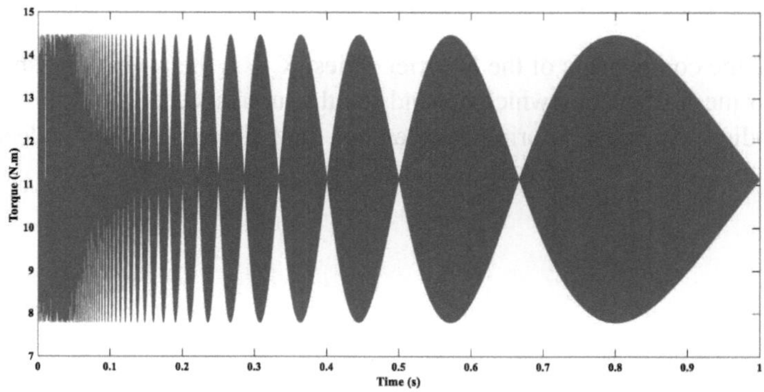

Fig. 3. Evolution of the wind torque

The wind torque has the same shape as the wind speed (Fig. 3).

The speed of the driving rotor depends on the wind velocity. It presents a balance until a referenced speed in order to improve wind turbine power. For this, the rotor speed is around a reference speed as presented in Fig. 4.

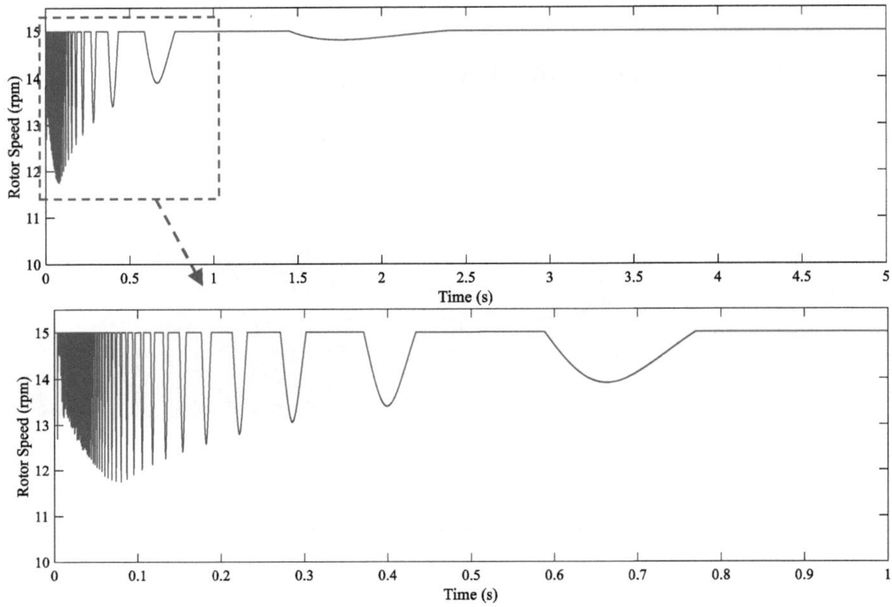

Fig. 4. Speed evolution of wind turbine rotor

The gear mesh stiffness in the two stages are sensitive to the speed variation of the input speed. It can be developed as a Fourier series (Lin and Parker, 2002):

$$K_{mi}(t) = K_{avi} + 2K_{ai}\sum\nolimits_{s=1}^{\infty}(a_s \sin(s\Omega_i t) + b_s \cos(s\Omega_i t)) \tag{7}$$

a_s and b_s are the components of the Fourrier series. K_{avi} is the average mesh stiffness, Ω_i is the gear mesh frequency which depends on the input speed.

Its periodicity increases during acceleration and decreases during deceleration (Fig. 5).

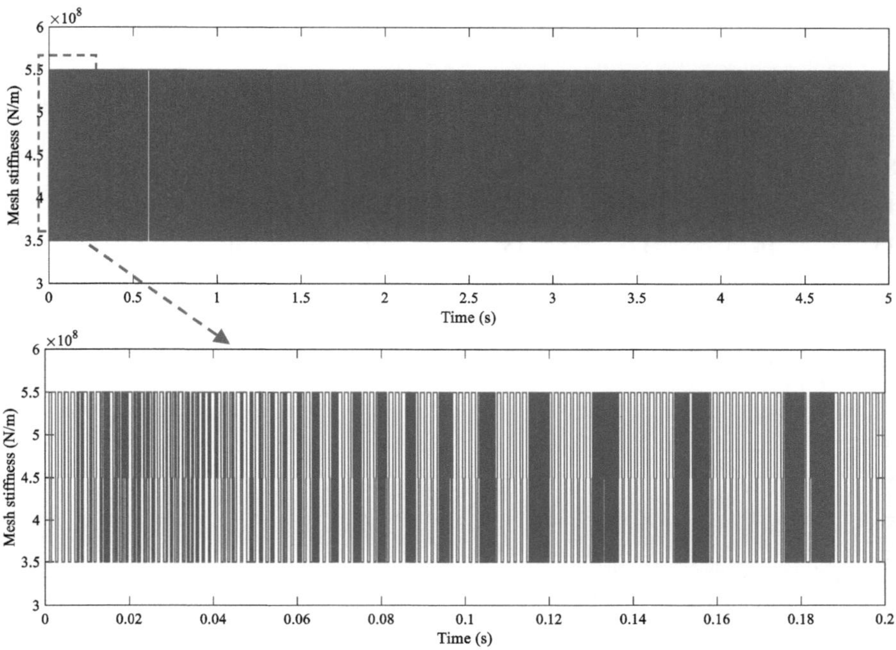

Fig. 5. Time evolution of gear mesh stiffness

4 Numerical Results

First, ω_i are computed:

$$det\left(-\omega_i^2 M + (K_s + K_{av}\right) = 0 \tag{8}$$

K_s is the structure matrix. K_{av} is the average matrix of $K(t)$.

The natural frequencies f_i are computed through:

$$\omega_i = 2\pi f_i \tag{9}$$

Computed natural frequencies are resumed in Table 2.

The acceleration signal at the sun bearing shows an oscillation around zero with amplitude modulation (Fig. 6).

In fact, the acceleration signal shows that the time variation of the vibration is sensitive to the wind torque and speed. A joint time-frequency plot is shown in Fig. 7.

Figure 7 shows vertical lines corresponding to the natural frequencies ($f_5 = 1271$ Hz, $f_{17} = 3223$ Hz and $f_{20} = 4804$ Hz). In addition, we observe square curves in the vertical direction. These curves correspond to the variation of the gear mesh frequencies of the planetary gear and the two stages spur gear. These curves present also a very important amplitude of vibration especially when they cross with the natural frequency ($f_{17} = 3223$ Hz).

Table 2. Natural frequencies (Hz)

1	**2**	**3**	**4**	**5**	**6**	**7**
30,46	900	975	975	1271	1525	1525
8	**9**	**10**	**11**	**12**	**13**	**14**
1677	1677	1778	2311	2330	2399	2637
15	**16**	**17**	**18**	**19**	**20**	**21**
2872	3071	3223	3882	4420	4804	4831
22	**23**	**24**	**25**	**26**	**27**	**28**
8831	17652	18218	121465	121516	121549	297046

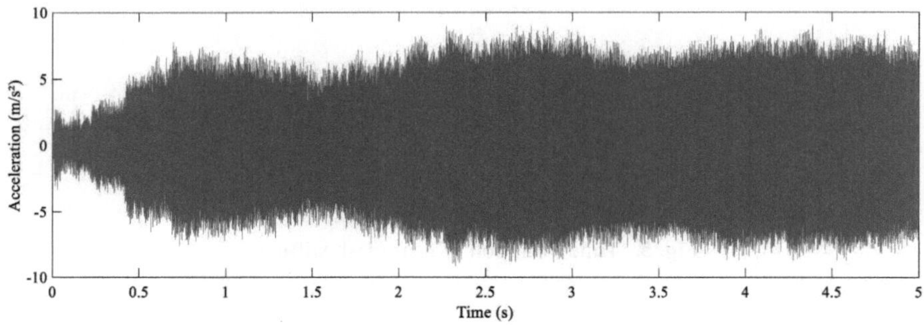

Fig. 6. Acceleration signal on the sun bearing

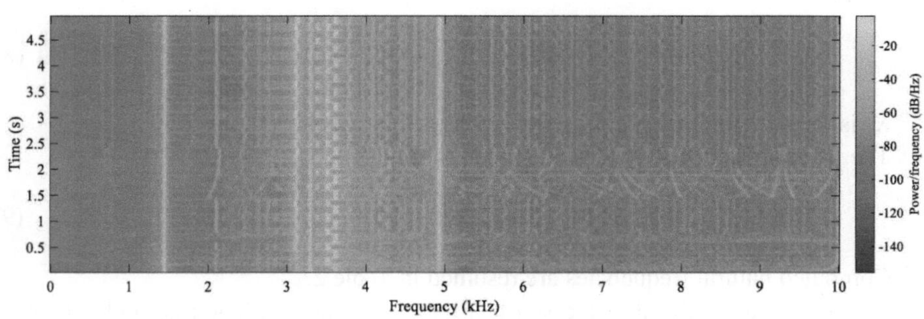

Fig. 7. Color map of the acceleration signal on the sun bearing

5 Conclusion

The aerodynamic effects on the dynamic behaviour of wind turbine gearboxes are studied in order to identify the frequency components that will be helpful to the wind turbine maintenance teams. A dynamic model of a WTG composed of a planetary gear, a two-stage spur gear and an elastic coupling is proposed. In this model, the aerodynamic effect is induced by the wind torque in the external forces and the wind speed.

The dynamic behaviour of the WTG is shown in the time domain and in the joint time-frequency domain, since wind turbines operate in transient conditions.

The dynamic vibration response of WTG is very sensitive to the aerodynamic effects and exhibits simultaneous frequency and amplitude modulations. In fact, the level of vibration increases especially when the gear mesh frequencies, which are variable, cross with the natural frequency.

6 Tribute

The authors would like to pay attribute to Professor Fakher CHAARI and appreciate his guidance and his contribution in this work. Pay to his soul.

Funding. This work is financed by the Tunisian-Mauritanian project n° "23MT1003", supported by the Tunisian and the Mauritanian Ministry of Higher Education through National Agency for Scientific Research and Innovation (ANRSI).

References

Abboudi, K., Walha, L., Driss, Y., Maatar, M., Fakhfakh, T., Haddar, M.: Dynamic behavior of a two-stage gear train used in a fixed-speed wind turbine. Mech. Mach. Theory **46**, 1888–1900 (2011)

Chaari, F., Abbes, M.S., Viadero, F., Fernandez, A., Haddar, M.: Analysis of planetary gear transmission in non-stationary operations. Front. Mech. Eng. **8**(1), 88–94 (2013)

Chakroun, A.E., et al.: Modal energetic analysis and dynamic response of worm gear drives with a new developed dynamic model. Comptes Rendus Mécanique. **349**(2), 241–258 (2021)

Dewangan, P., Parey, A., Hammami, A., Chaari, F., Haddar, M.: Damage detection in wind turbine gearbox using modal strain energy. Eng. Fail. Anal. **107**, 104228 (2020)

Dewangan, P., Parey, A., Hammami, A., Chaari, F., Haddar, M.: Dynamic characteristics of a wind turbine gearbox with amplitude modulation and gravity effect: theoretical and experimental investigation. Mech. Mach. Theory **167**, 104468 (2022)

Dewangan, P., Parey, A., Hammami, A., Chaari, F., Haddar, M.: Dynamic response computation of a wind turbine gearbox under variable speed conditions and its experimental validation. J. Vib. Eng. Technol. (2022). https://doi.org/10.1007/s42417-022-00782-8

Driss, Y., Hammami, A., Walha, L., Haddar, M.: Effects of gear mesh fluctuation and defaults on the dynamic behavior of two-stage straight bevel system. Mech. Mach. Theory **82**, 71–86 (2014)

Hadj Kacem, M., ELHami, A., Trabelsi, H., Walha, L., Haddar, M.: Considération de l'incertitude à l'étude du rendement d'un multiplicateur biétage d'une éolienne , Incert. Fiabilité Systèmes Multiphysiques (2021). https://doi.org/10.21494/ISTE.OP.2022.0778

Hammami, A., Fernandez, A., Viadero, F., Chaari, F., Haddar, M.: Dynamic behaviour of back to back planetary gear in run up and run down transient regimes. J. Mech. **31**(4), 481–491 (2015)

Hammami, A., Mbarek, A., Fernández, A., Chaari, F., Viadero, F., Haddar, M.: Dynamic behavior of the nonlinear planetary gear model in nonstationary conditions. Proc. Inst. Mech. Eng. C J. Mech. Eng. Sci. **235**(20), 4648–4662 (2021)

Hmida, A., Hammami, A., Khabou, M.T., Chaari, F., Haddar, M.: Effect of elastic coupling on the modal characteristics of spur gearbox system. Appl. Acoust. **144**, 71–84 (2019)

Kramer, E.: Dynamics of Rotors and Foundations, p. 1993. Springer-Verlag, New York (1993)

Lin, J., Parker, R.G.: Mesh stiffness variation instabilities in two-stage gear systems. J. Vib. Acoust. **124**(1), 68–76 (2002)

Mbarek, A., Fernández, A., Hammami, A., Chaari, F., Viadero, F., Haddar, M.: Early damage detection in planetary gear transmission in different operating conditions. Forschung Im Ingenieurwesen/Eng. Res. **86**(4), 861–874 (2022)

Tadeo, A.T., Cavalca, K.L., Brennan, M.J.: Dynamic characterization of a mechanical coupling for a rotating shaft. Proc. Inst. Mech. Eng. C J. Mech. Eng. Sci. **225**, 604–616 (2011)

Viadero, F., Fernández, A., Iglesias, M., de-Juan, A., Liaño, E., Serna, M.A.: Non-stationary dynamic analysis of a wind turbine power drivetrain: offshore considerations. Appl. Acoust. **77**, 204–211 (2014). https://doi.org/10.1016/j.apacoust.2013.10.006

Zghal, B., BelMabrouk, I., Walha, L., Abboudi, K., Haddar, M.: Analyses of dynamic behavior of vertical axis wind turbine in transient regime. Adv. Acoust. Vib. (2019)

Ziani, R., Hammami, A., Chaari, F., Felkaoui, A., Haddar, M.: Gear fault diagnosis under non-stationary operating mode based on EMD. TKEO Shock Detector Comptes Rendus Mécanique **347**(9), 663–675 (2019)

Zimroz, R., Urbanek, J., Barszcz, T., Bartelmus, W., Millioz, F., Martin, N.: Measurement of instantaneous shaft speed by advanced vibration signal processing – application to wind turbine gearbox. Metrol. Meas. Syst. XVIII. **4**, 701–712 (2011)

Dynamics of mechanical systems

Investigation of the Crack Impact on a Polymer Spur Gear Performance

Abderrahim Baccar[1,3,4](✉), Ahmed Hammami[1], Nuno Dourado[2], Ali Akrout[1], and Fakher Chaari[1]

[1] Laboratory of Mechanics, Modelling and Production, National School of Engineers of Sfax, University of Sfax, Sfax, Tunisia
abderrahim.baccar@gmail.com
[2] LABBELS–Associate Laboratory, Universidade Do Minho, 4800-058 SfaxGuimarães, Portugal
[3] National School of Engineers of Tunis, University of Tunis El Manar, Sfax, Tunisia
[4] Higher Institute of Technological Studies of Mahdia, Sfax, Tunisia

Abstract. Spur gears can be made using a variety of materials, including steel, aluminium, and polymer. Polymer gears are usually employed because they have different advantages over steel gears, such as great corrosion resistance, significant weight reduction, the ability to be used with or without lubrication, and especially a lower manufacturing price. However, failure modes prevent the use of polymer gears in all industrial applications. High temperatures and loads can threaten the polymer gear's life cycle by leading to the start of critical failure processes, such as cracks. Consequently, it is essential to examine how damage impacts the performance of polymer gears. Using the Extended Finite Element Method (XFEM) with a cohesive segment technique based on traction-separation laws, this study aims to examine the performance of a polyamide (PA12) spur gear. Specifically, this method will estimate the propagation of crack damage under different load magnitudes. In this study, a static analysis is conducted to evaluate the influence of various artificial fracture dimensions introduced into the tooth root on the mesh stiffness (MS), utilising a single-gear tooth model over one complete mesh cycle. Results show the influence of the different crack parameters on gear performance. A significant crack length and precise deviation lead to a reduction in the MS.

Keywords: Spur gear · Polymer · XFEM · Mesh Stiffness · Crack

1 Introduction

Since they are the most commonly employed gears, spur gears are widely investigated in order to enhance their performance and reach their potential. Various researchers, such as Mao et al. (2019) and Zhang et al. (2020), are considering the investigation of polymer gears due to their multiple advantages over steel gears. Zhang et al. (2020) identified the polymer gear failure modes, such as thermal bending, wear, and fatigue fractures. They demonstrate that all failure mechanisms are dependent on both life cycle and load magnitude and are critical to gears performance. Because of material fatigue, fractures

A. Hammami et al. (Eds.): MOSCOSSEE 2024, ACM 23, pp. 137–145, 2025.
https://doi.org/10.1007/978-3-032-00930-2_13

can affect gear teeth; thus, several studies, using various approaches, are focusing on the examination of fractured gears and the impact of cracks on the MS. Within this framework, Pandya and Parey (2013) utilised a 2-D finite element method (FEM) and taking into account the linear fracture mechanic theory to study the behaviour of cracks with different contact ratios. Different studies are conducted by the need to improve the computing of the MS, which is important in detecting gear defects. Mohammed et al. (2013) computed the MS of spur gears using a 2-D FEM. To assess the MS reduction due to a root fracture, Chaari et al. (2009) utilised both analytical and FEM approaches. For simplicity, they suggested a composed of set of teeth in three dimensions. Dogan et al. (2021) employ the FEM to investigate the fracture propagation for asymmetric and standard gears, as well as different rim thickness. Lewicki (1997) employs different methods to investigate the backup ratio (m_B) impact on the fracture path orientation. The existing literature indicates a notable gap in the investigation of the impact of fractures on the polymer gears performance.

The primary goal of this project is to explore the fracture propagation effect on a pair of gear tooth models. The suggested technique predicts the fracture propagation path without updating the mesh. This chapter will investigate the effect of artificial fracture characteristics on the polymer gears MS.

2 Mesh Stiffness Formulation

The simple pair spur gear MS depends on the tooth stiffness (TS) of driven and driving gears. In the pressure line direction, the load divided by the tooth deflection yields the TS. The TS for different gears can be determined using the equation below (Chaari et al. 2009),

$$K_{Tooth} = F/\Delta \tag{1}$$

where Δ and F indicate the tooth deflection on the pressure line direction and the total load. F is always perpendicular to the tangent line to the tooth profile.

The MS for a single tooth pair in mesh was established by Chaari et al. (2009) as:

$$K_m = K_D K_{Dg}/(K_D + K_{Dg}) \tag{2}$$

where K_m is the simple pair MS. K_D and K_{Dg} represent respectively the stiffnesses of the driving tooth and the driven tooth. The contact between the driven and driven teeth is not taken into consideration when calculating the simple pair MS. The calculation procedure of the MS can be executed for both normal and damages gears.

3 Finite Element Model

3.1 Three-Dimensional Model

To minimise the total amount of simulation time, the current model chosen as a part of the spur gear is presented in Fig. 1. This 3-D model is created by using the SOLID-WORKS software. This model is adopted for both driven and driving models. The gear characteristics that were employed in this analysis are shown in Table 1.

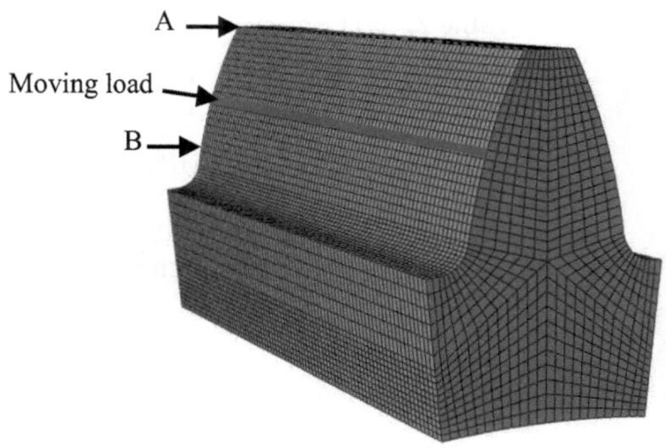

Fig. 1. Numerical tooth model.

Then, the model was uploaded into the Standard ABAQUS software. Since this mesh type is better suited for solid elasticity analysis (Chun Hung Lee. 2009), a linear hexahedron mesh and a structured grid were created for both driven and driving models (Fig. 1). Instead of simulating the contact between driven and driving teeth, the applied load is defined as pressure that moves on the surface |AB|.

Table 1. Spur gears feature values.

Parameters	Values
Module	$m = 2$
Teeth number	$Z_{Dg} = Z_D = 30$,
Pressure angle	$\alpha = 20°$
Face width	b = 17 mm
Whole depth	h = 4.5 mm
Rim thickness	r = 3.5 mm

For the Driven tooth model, the load goes from "A" to "B", whereas for the Driving model, the load moves from "B" to "A". Where the "A" and "B" edges represent the limits of the surface of contact (Fig. 1).

3.2 Material Properties

The polyamide 12 was chosen as the gear material for this FEM investigation. According to Baccar et al. (2023), this thermoplastic material is characterised by mechanical properties presented in Table 2. To explain the spread of damage, the cohesive properties of the PA12, which are displayed in Table 3, must be defined.

Table 2. PA12 properties.

Parameters	Values
Elastic modulus	$E = 1650$ MPa
Poisson's ratio	$\nu = 0.3$
Density	$M = 1.01$ g/cm^3

Table 3. PA12 damage properties (fracture mechanics test mode I).

Damage parameters	Values
Fracture toughness	$G_{IC} = 1.1$ N/mm
Interfacial Stiffness	$K_n = 1\text{E} + 06$ N/mm3
Cohesive strength	$\sigma_{nmax} = 16$ MPa
Relative displacement at damage onset	δ_{nmax} (mm)
Separation at failure	δ_{nf} (mm)

For this study, the PA12 polymer is considered an isotropic material. Consequently, the Mode I cohesive properties are attributed to others fracture mechanics modes (Mode II and Mode III) to facilitate the investigation of the mixed-mode crack propagation. The cohesive fracture mechanics mode I law is modelled as a linear-traction separation law (Fig. 2). The material elastic behaviour can be simulated thanks to the ascending segment (Fig. 2). After this domain, the beginning of damage occurs.

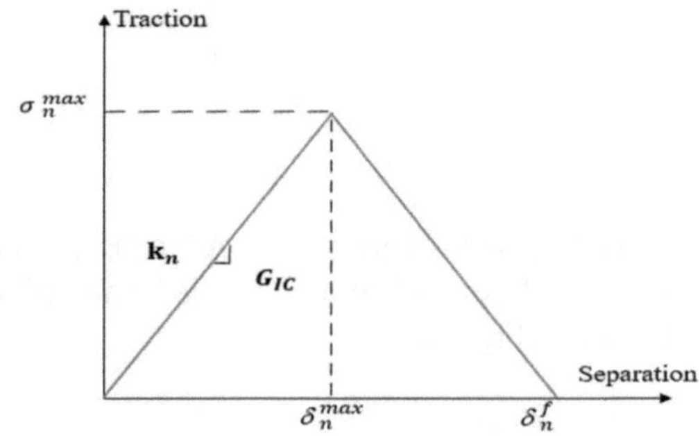

Fig. 2. Linear traction-separation law of the fracture mechanics test (mode I).

3.3 Crack Modelling

The XFEM is a computational method created by Belytschko (1999) to modelling ductile or brittle fracture. Unlike the standard methods, The XFEM utilises an approximate function for the displacement vector, incorporating additional enrichment functions. This made it possible to estimate fracture propagation without the necessity for re-meshing discontinuous surfaces, allowing cracks to travel along dependent routes, decreasing projection errors and computing costs. XFEM provides for an exact approximation of solutions, accommodating singularities, kinks, and other local non-smooth components.

Using ABAQUS, the XFEM can be used either with or without an initial introduced crack. To explore the crack features effect, we introduce an artificial one into the fillet root tooth (Fig. 3).

This fracture is characterised by a deviation angle β and an initial length a_0 (Table 3). This work begins by determining the simple pair MS for a faultless tooth. Then, as illustrated in Fig. 3, a crack is incorporated into the driven model. The MS is compared using the XFEM and taking into account variable fracture features and loads, as specified in Table 4.

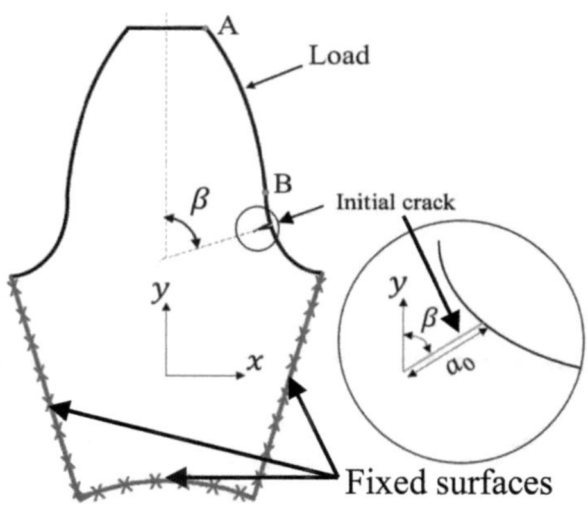

Fig. 3. Model shape.

Table 4. Parametric study.

Parameters	Values
Beginning crack angles deviation: β	80°/ 70°/ 60°/ 50°
Beginning crack lengths: a_0 (mm)	0/ 0.01/ 0.05/ 0.1
Loads (N)	100/150/200/250

4 Results

The MS and the TS for both driving and driven teeth, under 200N load, for one mesh cycle are shown in Fig. 4. The red curve corresponds to the driving TS, the blue curve refers to the driven TS and the black curve relates to the MS for regular teeth. The driven

TS tends to increase from the beginning until the very end of the mesh cycle (from 2017 N/mm to 9395 N/mm). Furthermore, the driving TS significantly reduces during the mesh cycle from 9406 N/mm to 1963 N/mm. The MS characterised by a maximum value of 2337.42 N/mm.

This leads us to conclude that the least TS value occurs when the load is on the corresponding tooth addendum and that the greatest stiffness corresponds to the instant when the pressure is on the bottom of the tooth. The tooth is exposed to extreme stress localised in the root region while transferring engine torque, which causes the crack to form. The MS is displayed in the next figures (Figs. 5, 6 and 7) for initial artificial fracture lengths a_0 of 0.01, 0.05 and 0.1 mm respectively. In spite of the fact that the length of the initial crack was less than or equivalent to 0.01 mm, we came to the conclusion that the MS did not experience any significant impact.

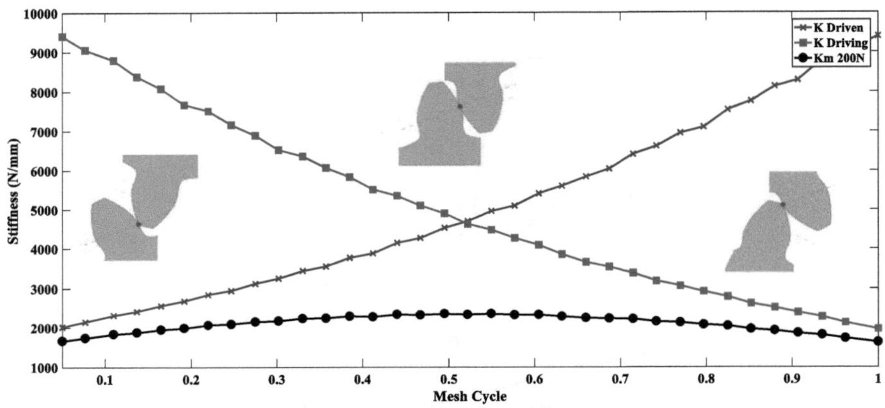

Fig. 4. MS and tooth stiffness for undamaged teeth (200N load).

Additionally, regardless of the initial angle or the load value that was applied, seeing as how all curves appear to be similar (Fig. 5). Furthermore, when the load value remains the same and the first crack length a_0 is approximately 0.05 mm or even 0.1 mm, a drop in stiffness is observed for each curve at the initiation of the mesh cycle. This reduction occurs when the initial fracture angle β is equal to either 70° or 80°, as depicted in Figs. 6 and 7. The crack that has a short starting length does not have a significant impact on the MS curve, as the situation actually stands. The angle β has a considerable effect observed for each length a_0 (0.01 mm, 0.05 mm, and 0.1 mm), regardless of the load magnitude (100, 150, 200, and 250 N). The mean MS values for all cases with an initial angle β of 60° and 50° are quite similar. However, when β is equal to 80° or 70°, a significant MS reduction occurs. The mean MS for 50° and 60°, for instance, is 2090.87 N/mm and 2091.13 N/mm, respectively, with $a_0 = 0.05$ mm under 200 N. This result is nearly identical to the mean MS obtained with the intact model, which is 2090.86 N/mm. For fractured models with angles β of 70° and 80°, the mean MS decreases to 2022.96 N/mm and 2022.07 N/mm, respectively.

Another essential consideration concerns the crack length ratio. This ratio is calculated by dividing the ultimate fracture length by the length of the crack ligament. Results demonstrate that with a tooth thickness of 3.89 mm at the starting fracture site, the crack spreads across the tooth width, changing direction along the tooth thickness.

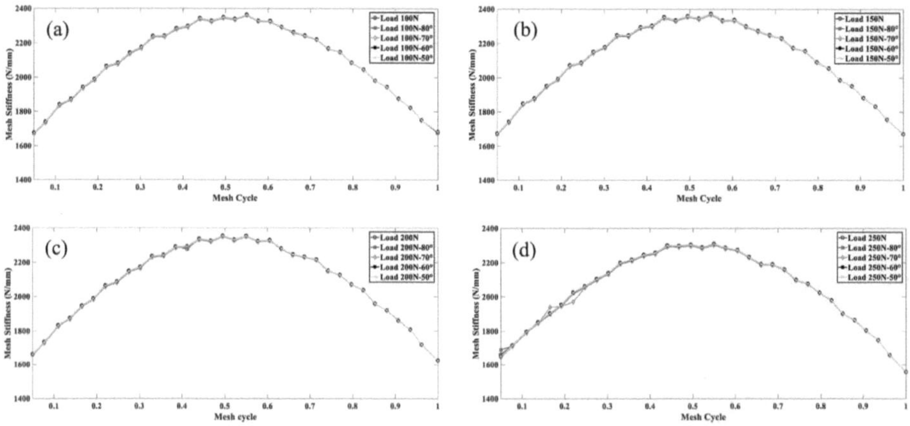

Fig. 5. MS with $a_0 = 0.01$ mm and different crack angles deviation (50°, 60°, 70° and 80°) (a) 100N (b) 150N (c) 200N (d) 250N.

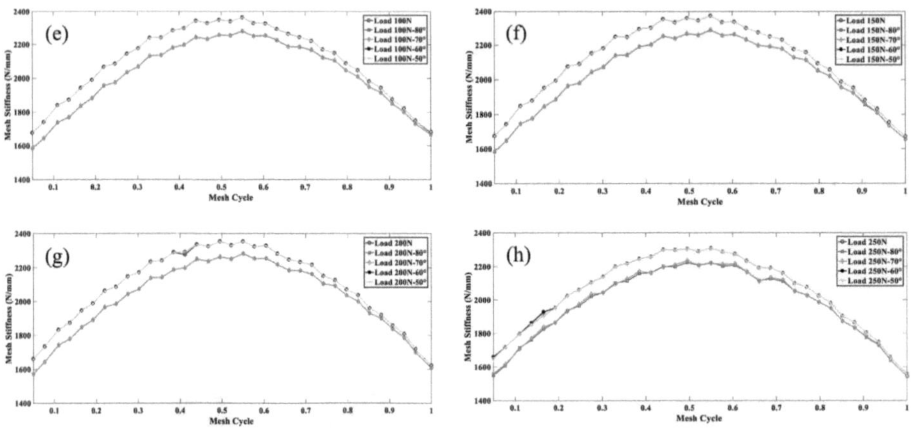

Fig. 6. MS with $a_0 = 0.05$ mm and different crack angles deviation (50°, 60°, 70° and 80°) (e) 100N (f) 150N (g) 200N (h) 250N.

Figure 8 illustrates the configuration of a crack originating from an initial singularity measuring 0.1 mm, with initial orientations of 80°.

In the instance of a 0.1 mm initial fracture, the crack has expanded only for 70° and 80° model cases. For β angles of 60° and 50°, the same crack length ratio (0.025) has been registered, with no crack propagation seen. Models with β angles of 80° and 70° have crack length ratios of 0.09 and 0.08, respectively, indicating that the fracture is able to propagate (Fig. 9).

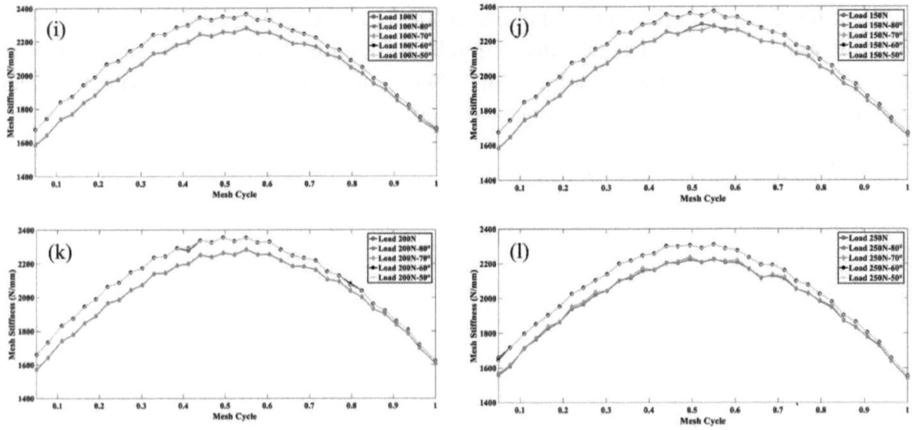

Fig. 7. MS with $a_0 = 0.1$ mm and different crack angles deviation (50°, 60°, 70° and 80°) (i) 100N (j) 150N (k) 200N (l) 250N.

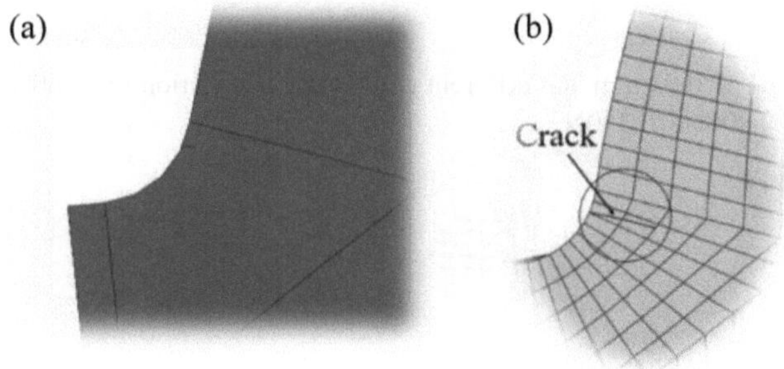

Fig. 8. Cracks growth under 250 N load (a) Initial crack configuration ($a_0 = 0.1$ mm; $\beta = 80°$) (b) Crack propagation ($a_0 = 0.1$ mm; $\beta = 80°$).

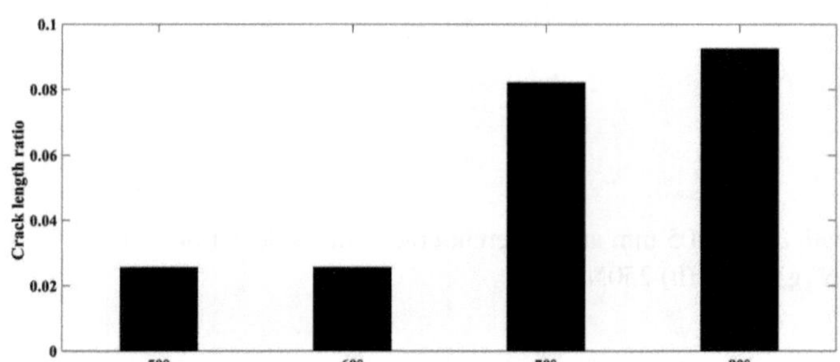

Fig. 9. Crack length ratio under 250 N ($a_0 = 0.1$ mm; $\beta = [\,50°, 60°, 70°, 80°]$).

Lewicki (1997) discovered that the fracture spreads across the tooth width for models with $m_B \geq 0.5$ (rim thickness divided by the tooth depth). For this simulation ($m_B = 0.77$), the fracture progressed into the tooth thickness, consistent with Lewicki's findings.

5 Conclusion

This work employs the XFEM to investigate the impact of different fracture features on the MS for polymer spur gears. The XFEM allows the perfect numerical inspection of the fracture propagation and its impact on the MS. The crack, influenced by the stress and originating from the contact side at the tooth root, may affect the toughness of the tooth. Damage is indicated by a decrease in the MS curve. A maximum stiffness reduction is observable at the initiation of the mesh cycle, where the cracked tooth (driven) is just engaged. We find that the crack geometrical features (a_0 and β) have a significant impact on the fracture propagation over the cycle. The observation that, with $m_B = 0.77$, the fracture propagates across the tooth thickness agrees with Lewicki's results (1997).

6 Tribute

Authors would like to pay attribute to Professor Fakher CHAARI and appreciate his guidance and his contribution in this work. Pay to his soul.

Acknowledgements. This work is supported by Tunisian-Portuguese project collaboration between Mechanics, Modelling and Production laboratory (LA2MP) and LABBELS –Associate Laboratory.

References

Baccar, et al.: Determination of the crack propagation effect on the mesh stiffness for a polymer spur gear tooth using the extended finite element method. J. Braz. Soc. Mech. Sci. Eng. **45**, 286 (2023)

Belytschko: Elastic crack growth in finite elements with minimal remeshing. Int. J. Numer. Methods Eng. (1999)

Chaari, F., et al.: Analytical modelling of spur gear tooth crack and influence on gearmesh stiffness. Eur. J. Mech. A/Solids **28**(3), 461–468 (2009)

Chun Hung Lee: Non-Linear Contact Analysis of Meshing Gears. Thesis Presented to The Faculty of California Polytechnic State University, San Luis Obispo (2009)

Do˘gan, et al.: Effects of rim thickness and drive side pressure angle on gear tooth root stress and fatigue crack propagation life. Eng. Fail. Anal. (2021)

Lewicki: Effect of rim thickness on gear crack propagation path. ASME (1997)

Mao, K., et al.: The wear resistance improvement of fibre reinforced polymer composite gears. Wear **426–427**, 1033–1039 (2019). https://doi.org/10.1016/j.wear.2018.12.043

Mohammed, O.D., et al.: Improving mesh stiffness calculation of cracked gears for the purpose of vibration-based fault analysis. Eng. Fail. Anal. **34**, 235–251 (2013). https://doi.org/10.1016/j.engfailanal.2013.08.008

Pandya, Y., Parey, A.: Failure path based modified gear mesh stiffness for spur gear pair with tooth root crack. Eng. Fail. Anal. **27**, 286–296 (2013). https://doi.org/10.1016/j.engfailanal.2012.08.015

Zhang, Y., et al.: A physical investigation of wear and thermal characteristics of 3D printed nylon spur gears. Tribol. Int. **141**, 105953 (2020). https://doi.org/10.1016/j.triboint.2019.105953

Investigation of the Influence of Viscoelastic Layer Thickness and Stacking Configurations on the Dynamic Behavior of Flax/PLA Biocomposites

Hajer Daoud[1,2](✉), Firas Meddeb[1,3], Abderrahim ElMahi[3], Jean-Luc Rebiere[3], Mohamed Amine Ben Souf[1], and Mohamed Haddar[1]

[1] Laboratory of Mechanics Modeling and Production (LA2MP), National School of Engineers of Sfax, University of Sfax, BP N° 1173, 3038 Sfax, Tunisia
doud.hajer@yahoo.fr, {firas.meddeb, mohamed-amine.bensouf}@enis.tn

[2] Institute of Higher Studies in Applied Sciences and Technology of Kairouan, Kairouan, Tunisia

[3] Acoustics Laboratory of the University of Le Mans (LAUM), UMR CNRS 6613, Le Mans University, 72085 Le Mans, France
{abderrahim.elmahi,jean-luc.rebiere}@univ-lemans.fr

Abstract. In this paper, experimental vibration analyses were carried out on biocomposite materials, both with and without an interleaved natural viscoelastic layer. The composites consisted of an exterior Flax/PLA layer and interleaved rubber layers (TPU) of varying thicknesses and positions. The different configurations were manufactured using advanced 3D printing technology. The experimental tests were performed on beam specimens with different stacking sequences, under a clamped-free configuration, utilizing an impulse technique in flexural vibration. This setup allowed for the precise measurement of the composite's vibrational behavior under varying conditions. To assess the dynamic properties of the composites, resonance vibration experiments were conducted. These experiments were aimed at determining key dynamic parameters, including natural frequencies and modal loss factors for each bending mode. The influence of the viscoelastic layer's thickness, number, and positioning was carefully examined to understand its impact on the damping behavior of the composites. The results highlight the significant role that the viscoelastic layer plays in enhancing the damping performance of the materials, demonstrating how variations in layer configuration can lead to improved dynamic behavior. This improvement is primarily due to the energy dissipation properties of the viscoelastic layer, which effectively absorbs and dissipates vibrational energy. This study provides valuable insights for the design and optimization of composite materials with enhanced vibration damping capabilities.

Keywords: Biocomposite materials Viscoelastic layers · Vibration analysis · Damping performance · 3D printing

A. Hammami et al. (Eds.): MOSCOSSEE 2024, ACM 23, pp. 146–153, 2025.
https://doi.org/10.1007/978-3-032-00930-2_14

1 Introduction

Biocomposites, composed of natural fibers and biodegradable polymer matrixes, have garnered significant atention in recent years as sustainable solutions for various industrial applicaitons. These materials combine lightweight properties, adequate mechanical performance, and low environmental impact, making them particulary attractive for use in the automotive, aerospace, and construction industries [1–3]. Among the promissing combinations, flax fiber and PLA-based composites stand out due to their biodegradability and compatability with advanced manufacturing techniques, such as 3D printing [4, 5].

Despite their many advantages, the aplication of biocomposites in dynamic eviroments remains challenging. One of the main limitations is their inablity to dissipate energy efficiently when subjected to vibrations. Dynamic loads can lead to premature material degredation, reduced lifespan, and performance instability, particularly in scenarios where damping is critical [6, 7]. These layers not only improves energy dissipation but also alow for ameliorating properties through the careful selection of material thickness and positions [8, 9].

The adoption of advanced manufactruing techniques such as 3D printing has further revolutionized the development of composite materials. This technology could control the stacking sequences, thickness distributions, and orientation of different layers, wich allow to explore inovative designs for improving material performances [10]. Moreover, 3D printing presents a good solution to the integration of viscoelastic layers into biocomposites, providing a unique advantage in terms of customization, interface and repeatability during fabrication [11].

This study investigtaes the effect of incorporating thermoplastic polyurethane (TPU) interleaved layers in Flax/PLA biocomposites manufactured via 3D printing. The primary objective is to asses the impact of the thickness, number, and position of these layers in the dynamic properties such as natural frequencies, damping factors. Using varied stacking configurations, this research provides valuable insights into the optimization of bio-based composites for applications where vibration control is essential.

2 Materials and Methods

In this study, beams are composed with layers of Flax/PLA and layers of natural rubber having different thicknesses. The outer layer was produced by PLA (polylactic acid) and Flax fiber filaments, while the interleaved layers were made from natural thermoplastic polyurethane rubber (TPU).

The rubber layers were incorporated to enhance the damping performance of the composites. Three different thicknesses were used: 1 mm, 1.5 mm, and 2 mm. Each thickness was applied in configurations with either one or two interleaved layers, looking for the optimization of damping properties through variation in layer number and thickness.

Figure 1 illustrates the fabricated composite beams, highlighting different configurations of rubber layer thicknesses and stacking sequences employed in this study.

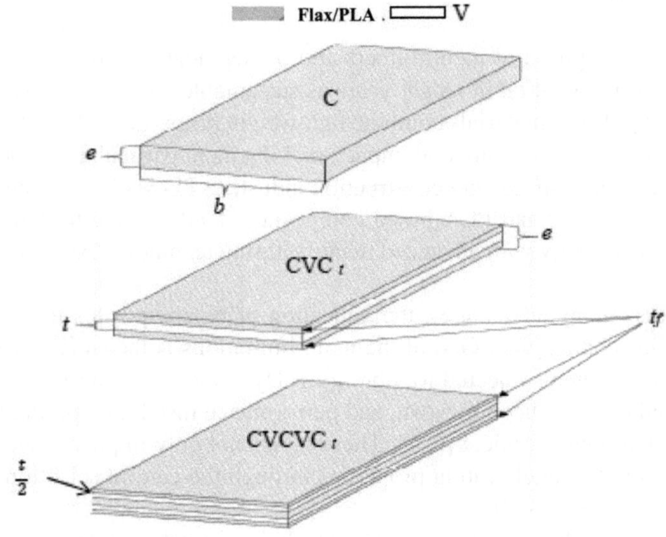

Fig. 1. Composite configurations: PLA/Flax_Viscoelastic layer (TPU)

The composite beams were manufactured using 3D printing technique. The printed samples were obtained by alternating layers of PLA/Flax and rubber. Due to the used manufacturing process, the printed samples were isotropic in each layer. The layer stacking was labeled as follows: Without Viscoelastic Layer labeled (C); with one viscoelastic layer labeled (CVC_t) and with two viscoelastic layers ($CVCVC_t$) where t is is the thickness of the viscoelastic layer (1, 1.5, or 2 mm).

Table 1 presents the mechanical properties of Flax/PLA and rubber used in this work. These values were obtained from previous studies realised by Firas et al. [12].

Table 1. Mechanical Properties of Flax/PLA and Rubber [12]

Material	Young's Modulus (MPa)	Density (Kg/m^3)	Poisson's ratio
Flax/PlA	2720 ± 10	1070 ± 15	0.3 ± 0.01
Rubber	12.5 ± 1	1190 ± 15	0.5 ± 0.04

3 Experimental Setup

The experimental setup was used to evaluate the dynamic response of the obtained beams under impact excitation. A schematic representation of the setup is presented in Fig. 2. The excitation was provided by a hammer impact. The resulting vibrations were detected using a laser vibrometer, which offered high precision in measuring the displacement and velocity of the beam's surface. For the analysis of the frequency response, the (−3 dB) bandwidth method was employed. This technique involves identifying the frequency range where the system's response amplitude remain within 3 decibels of its peak value, providing insights into the damping characteristics and resonant frequencies of the composite beams. This experimental approach enabled a comprehensive assessment of the mechanical and vibratory properties of composite beams, facilitating the optimization of their design for enhanced performance.

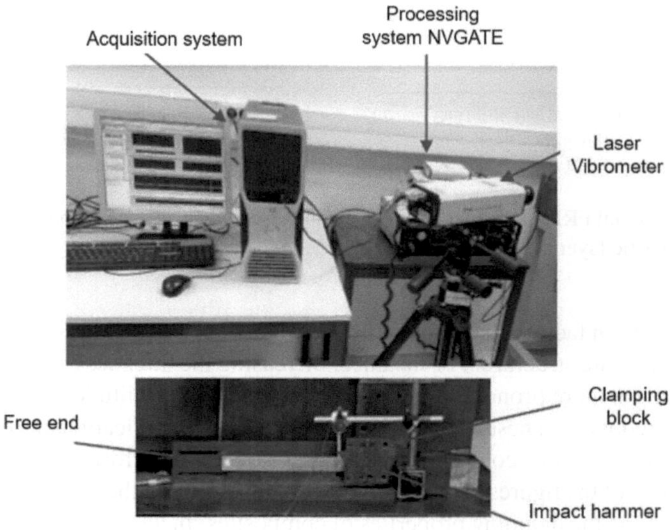

Fig. 2. Experimental setup.

4 Results and Discussion

The evolution of the amplitude of the frequency response function (FRF) as a function of frequency is presented in Fig. 3. The results indicated that, for all studied composites, incorporating a viscoelastic layer reduce vibration amplitude and shift resonance peaks to lower frequencies. These findings align with previous researchers investigated by Firas et al. [13] on the damping effects of viscoelastic layers in composite structures. Furthermore, incorporating a viscoelastic layer improve the material's damping properties, which result in lower peak amplitudes, shift resonance frequencies, and broaden resonance peaks.

Figure 4 illustrates the evolution of the loss factor (η) as a function of vibration frequency for each composite studied. The incorporation of a viscoelastic layer into composite materials significantly enhances their damping properties. This enhancement is primarily due to the viscoelastic layer's ability to dissipate vibrational energy through shear deformation, effectively converting mechanical energy into heat. As a result, composites with viscoelastic layers exhibit reduced vibration amplitudes and altered resonance frequencies compared to their non-viscoelastic counterparts. The extent of these

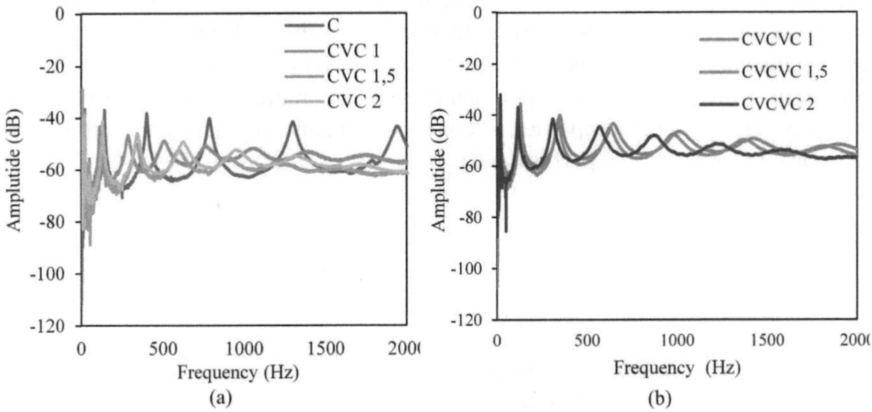

Fig. 3. Experimental FRFs: a) Effect of inserting one viscoelastic layer; b) Composites interleaving two viscoelastic layers.

changes depends on factors such as the thickness and placement of the viscoelastic layer within the composite structure. For instance, increasing the thickness of the viscoelastic layer can lead to a more pronounced reduction in vibration amplitude and a greater shift in resonance frequency. These modifications are crucial for applications where vibration control is essential, as they contribute to improved structural performance and longevity.

The analysis of the figures provides significant insights into the impact of adding viscoelastic layers on the damping properties of composites. In terms of damping, incorporating a single viscoelastic layer results in higher damping levels compared to composites without such a layer, regardless of the layer's thickness. Notably, composites with a single viscoelastic layer exhibit better damping than those with two interleaved viscoelastic layers, especially at a thickness of 1 mm. These findings highlight the importance of the continuity of the viscoelastic layer for effective vibrational energy absorption. Overall, these results underscore the significance of layered composite design in optimizing damping performance, considering factors such as the arrangement of viscoelastic layers, layer continuity, and the number of interfaces and discontinuities within the material.

The Evolution of damping factors across the first four vibration modes and for the different studied composites are presented in Fig. 5. The results reveals that incorporating a single 1 mm (CVC1) or 1.5 mm (CVC1.5) viscoelastic layer significantly enhances damping. For the first mode, η increases from 2.16% in the base composite to 4.7% and 6%, respectively. Adding two viscoelastic layers (CVCVCt) yields variable results: a 1 mm and 1.5 mm combination (CVCVC1) reaches 3.97%, while two 1 mm layers (CVCVC2) result in 3%. This indicates that both the thickness and number of layers affect damping, with greater thickness generally improving efficiency.

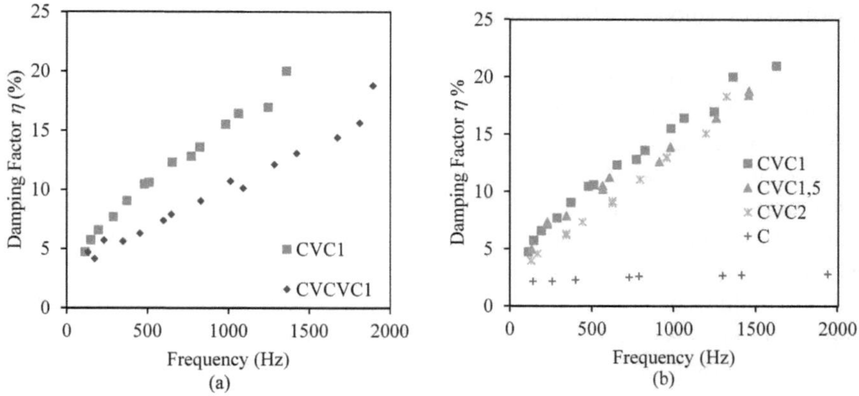

Fig. 4. Evolution of the loss factor (η) as a function of frequency: a) influence of stacking sequence and b) influence of viscoelastic layer thickness.

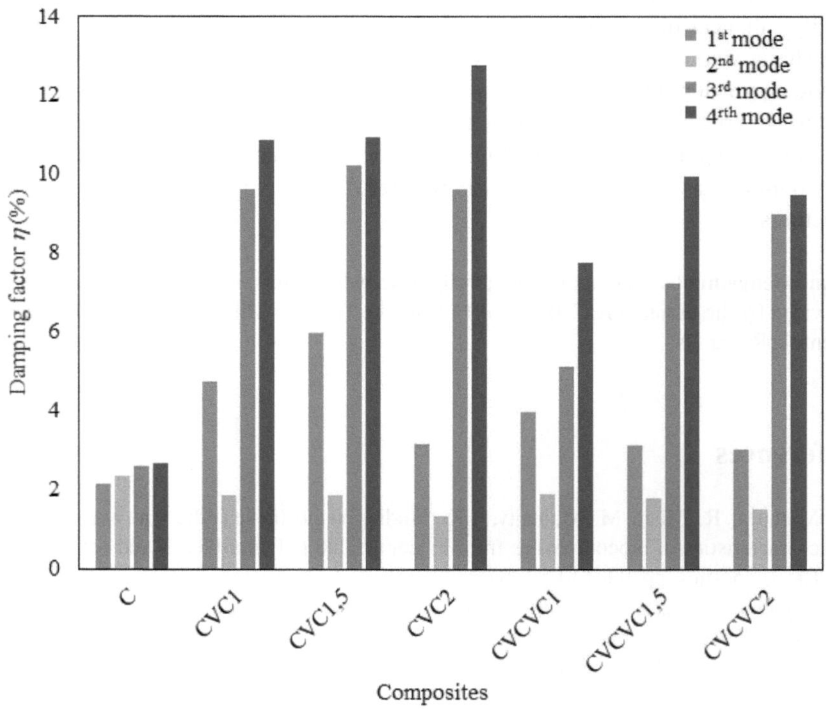

Fig. 5. Evolution of damping factors for different composites

In the second mode, the base material exhibits a damping coefficient of approximately 2.3%. A 1 mm layer (CVC1) raises η to 6.7%, and a 1.5 mm layer (CVC1.5) to 6.9%. However, two viscoelastic layers reduce damping effectiveness compared to a single layer, with η at 4.2% for CVC1 and 5.3% for a total thickness of 1.5 mm in CVCVCt.

For the third and fourth modes, a 1 mm viscoelastic layer significantly increases the loss factor to 9.6% and 10.8%, respectively. Increasing the layer thickness to 1.5 mm further raises η to 10.1% and 10.9%. This trend suggests that thicker layers consistently enhance damping. However, adding a second viscoelastic layer (CVCVC1) decreases the loss factor, possibly due to complex interactions between the layers.

These findings demonstrate that integrating viscoelastic layers effectively improves damping in Lin/PLA composites across all vibration modes. The layer thickness significantly influences η, with thicker layers dissipating more energy and resulting in higher loss factors. While adding a second viscoelastic layer can also enhance damping, its effect is more complex and depends on the thickness of both layers. In some cases, adding a second layer may reduce the loss factor.

5 Conclusion

In conclusion, the integration of viscoelastic layers into composite materials significantlly enhances their damping properties across various vibration modes. The addition of a single viscoelastic layer notably incerases the damping factor, with thicker layers providing greater improvements. However, incorporating multiple viscoelastic layers can lead to variable effects on damping eficiency, influenced by factors such as layer thickness and arrangement. These findings underscore the importance of carefully selecting and configuring viscoelastic layers to optimize the damping performance of composite structures.

Acknowledgements. The authors gratefully acknowledge the assistance and financial support pro-vided by the project PEJC2023-D6P12 from the Tunisian Ministry of Higher Education and Scientific Research.

References

1. Muthuraj, R., Misra, M., Mohanty, A.K.: Studies on mechanical, thermal, and morphological characteristics of biocomposites from biodegradable polymer blends and natural fibers. In: Biocomposites, pp. 93–140. Elsevier (2015)
2. Spyridonos, E., et al.: LeichtPRO-profiles: development and validation of novel linear biocomposite structural components fabricated from pultruded natural flax fibres with plant-based resin for sustainable architectural applications. Mater. Res. Exp. **11**(7), 075308 (2024). https://doi.org/10.1088/2053-1591/ad6536
3. Hammami, A., Heyns, P.S., Schmidt, S., Chaari, F., Abbes, M.S., Haddar, M. (eds.): Modelling and Simulation of Complex Systems for Sustainable Energy Efficiency: Contributions to the First International Workshop on Modelling and Simulation of Complex Systems for Sustainable Energy Efficiency, MOSCOSSEE'2021, February 25–26, 2021. Springer International Publishing (2022)

4. Essassi, K., et al.: Damping analysis and failure mechanism of 3D printed bio-based sandwich with auxetic core under bending fatigue loading. J. Renew. Mater. **9**(3), 569–584 (2021). https://doi.org/10.32604/jrm.2021.012253
5. Hamrouni, A., et al.: Experimental and finite element analyses of a 3D printed sandwich with an auxetic or non-auxetic core. J. Sandwich Struct. Mater. **25**(4), 426–444 (2023). https://doi.org/10.1177/10996362231151454
6. Daoud, H., et al.: Numerical and experimental characterization of the dynamic properties of flax fiber reinforced composites. Int. J. Appl. Mech. **08**(05), 1650068 (2016). https://doi.org/10.1142/s175882511650068x
7. Ramalingam, R., et al.: Aging effects on free vibration and damping characteristics of polymer-based biocomposites: a review. Polym. Compos. **43**(6), 3890–3901 (2022). https://doi.org/10.1002/pc.26664
8. Zhang, G., Zheng, H., Zhu, X.: Optimization of composite plates with viscoelastic damping layer for high sound transmission loss under stiffness and strength constraints. Compos. Struct. **306**, 116563 (2023). https://doi.org/10.1016/j.compstruct.2022.116563
9. Daoud, H., et al.: Characterization of the vibrational behaviour of flax fibre reinforced composites with an interleaved natural viscoelastic layer. Appl. Acoust. **128**, 23–31 (2017). https://doi.org/10.1016/j.apacoust.2016.12.005
10. Asai, G., et al.: Ecodesign of 3D volumetric fiber-composite structures with topology optimization. Compos. Part A Appl. Sci. Manuf.Part A Appl. Sci. Manuf. **190**, 108615 (2025). https://doi.org/10.1016/j.compositesa.2024.108615
11. Hamrouni, A., et al.: Poisson's ratio influence on structural dynamic properties of 3D printed bio-based sandwich with an anti-trichiral core. J. Vib. Control JVC **30**(13–14), 2803–2815 (2024). https://doi.org/10.1177/10775463231185914
12. Meddeb, F., et al.: Effects of temperature and layer thickness on the static and dynamic behaviors of biocomposite materials with interleaved viscoelastic layers. Int. J. Appl. Mech. (2024). https://doi.org/10.1142/s1758825124500431
13. Meddeb, F., et al.: Experimental investigation on the static and dynamic behavior of passive controlled bio composite manufactured via 3D printing technique. Mech. Compos. Mater. **60**(3), 487–500 (2024). https://doi.org/10.1007/s11029-024-10205-2

Enhancing Lifespan Predictions for Shaft Bearings Using Feed-Forward Neural Networks

Rawnak Omar[1,2(✉)], Chaima Chikhaoui[1], Fathi Djemal[1,2], and Hassen Trabelsi[1,3]

[1] Laboratory of Mechanics, Modeling and Production (LA2MP), National School of Engineers of Sfax, University of Sfax, BP1173, 3038 Sfax, Tunisia
rawnakomar1997@gmail.com

[2] National Engineering School of Gabes, University of Gabès, Av Omar Ibn El Khattab, Zrig Eddakhlania, 6029 Gabès, Tunisia

[3] Higher Institute of Industrial Systems of Gabes, University of Gabès, Cité erriadh 6072 zrig, Gabès, Tunisia

Abstract. Accurately estimating a machine's Remaining Useful Life (RUL) is essential for improving maintenance efficiency and ensuring operational reliability. This study proposes an Artificial Neural Network (ANN) to enhance RUL prediction for bearing failures. The model processes both historical and real-time data to improve accuracy. The ANN takes root mean square, kurtosis, time, and adjusted Weibull failure rate as inputs. These features capture essential degradation patterns. To reduce noise, the model uses standardized operational duration percentages as output. This approach improves stability and prediction reliability.A Feedforward Neural Network optimized with the Levenberg-Marquardt algorithm is implemented. Training and validation assess its performance. The model architecture includes two hidden nodes, ensuring a balance between complexity and efficiency. Results show low training and validation errors, confirming the network's accuracy.

The proposed method provides a robust tool for predictive maintenance. By detecting early degradation, it helps optimize maintenance schedules and reduce unexpected failures. Compared to traditional approaches, this model offers better adaptability to non-linear degradation trends. Future work will focus on extending validation to different bearing types and operating conditions. This research contributes to the advancement of predictive maintenance strategies by integrating advanced machine learning techniques for more precise RUL estimation.

Keywords: Remaining Useful Life · Artificial Neural Network · training · prediction

1 Introduction

With the integration of internet-connected sensors, production equipment can be monitored remotely and in real time. This allows maintenance teams to address potential issues before they occur, reducing costs through preventive maintenance. Vibration analysis is widely used in predictive maintenance due to its ability to extract valuable information from vibration signals (Popescu et al., 2022; Hammami et al., Hammami et al. 2020).

A. Hammami et al. (Eds.): MOSCOSSEE 2024, ACM 23, pp. 154–167, 2025.
https://doi.org/10.1007/978-3-032-00930-2_15

Accurately identifying a suitable metric for monitoring bearing degradation is crucial for predicting the Remaining Useful Life (RUL) (Huang et al., 2007). An effective indicator must detect early-stage degradation and align with the degradation process (Pan et al., 2010). As bearing fatigue progresses, spalls form on the race surface, generating defect frequencies. The intensity of these frequencies correlates with spall size (William and Hoffman 2011). Therefore, the intensity of defect frequencies, combined with its first six harmonics, is used as a key metric for estimating bearing RUL (Gebraeel et al., 2004).

Accurate RUL prediction is a critical component of Predictive Health Management (PHM) since maintenance decisions rely on it (Si et al., 2013). Various measurements are considered for RUL forecasting, including vibration signals, temperature readings, oil analysis, and pressure measurements. Among these, vibration analysis is the most commonly used due to its simplicity and ability to provide comprehensive insights into bearing condition (Pham et al., 2010).

Prognostic methods are generally classified into three categories: physics-based models, data-driven models, and hybrid approaches. Physics-based models rely on crack growth analysis, integrating mechanical principles, flaw propagation dynamics, and Condition-Based Monitoring (CBM) data. For example, Li et al. (1999, 2000) developed a model that estimates bearing RUL by analyzing defect growth and fatigue crack progression. Similarly, Li and Lee (2005) used Paris law to study gear fatigue crack growth, validating their model with experimental data.

Tian (2009) introduced an Artificial Neural Network (ANN) to improve the accuracy of pump bearing RUL prediction. The model used multiple condition-monitoring metrics as inputs and pump bearing life percentage as the output.

Fordal et al. (2023) highlighted the effectiveness of Artificial Neural Networks (ANN) in predictive maintenance, particularly for high-reliability industrial systems. ANN processes machine data for fault detection and risk assessment, enhancing maintenance efficiency and reducing downtime.

Khalil and Rostam (2024) investigated the application of machine learning for predictive maintenance in rotating machinery, highlighting the effectiveness of algorithms for early fault detection. Their study demonstrates that these techniques enhance equipment reliability by reducing unexpected failures and maintenance costs.

In this study, we propose an ANN-based approach for estimating the RUL of failed bearings. The model takes as input time-based data, adjusted Weibull parameters, root mean square (RMS) values, and kurtosis failure rates derived from both current and past observations. The normalized life percentage is used as the output variable to reduce noise and enhance prediction accuracy. A Feedforward Neural Network optimized with the Levenberg-Marquardt algorithm is implemented for this purpose.

To ensure accurate RUL estimation, signal preprocessing is necessary before feeding data into the ANN. The next section presents the preprocessing techniques applied to vibration signals, focusing on feature extraction and noise reduction.

2 Signal Pretreatment

Bearings play a crucial role in wind turbines. Their failures are often caused by improper installation, insufficient lubrication, misalignment, excessive tolerance, or poor handling. Effective fault monitoring is essential for detecting early signs of degradation, preventing severe damage, and minimizing safety risks.

Various techniques are used to detect and analyze bearing failures, including vibration analysis and acoustic emission. In this study, we focus on vibratory signals collected from wind turbine bearings. The RMS, kurtosis, and skewness values are extracted as fault indicators.

To further characterize bearing degradation, the Weibull failure rate is used as a proxy for actual measurements. This failure rate provides a reliable metric for assessing machine condition and component reliability. The vibratory signals used in this research are sourced from the Center for Intelligent Maintenance Systems (IMS) dataset (Lee et al., 2020).

As shown in Fig. 1, consists of a test bench designed to evaluate bearing performance under controlled conditions. It includes four Rexnord ZA-2115 double-row bearings mounted on a rotating shaft. The shaft is driven by an AC motor via a friction belt drive, maintaining a constant rotational speed of 2000 rpm.

To simulate operational loads, a radial force of 6,000 lb is applied using a calibrated spring mechanism. Vibration data is captured with high precision using PCB 353B33 quartz ICP accelerometers. These accelerometers are securely mounted on each bearing housing to track vibration signals accurately.

Measurements are taken every 10 min, with each recording lasting 1 s. A high sampling frequency of 20 kHz generates 20,480 data points per measurement. This setup provides high temporal resolution for analyzing bearing dynamics and failure characteristics.

Lubrication is maintained through an oil pump. The presence of debris on the magnetic plug in the oil return line indicates bearing degradation. The test concludes when the accumulated debris surpasses a predefined threshold (Qiu et al., 2006).

The extracted features from the vibration signals serve as key inputs for the ANN model. In the next section, we describe the structure and training of the neural network for RUL prediction.

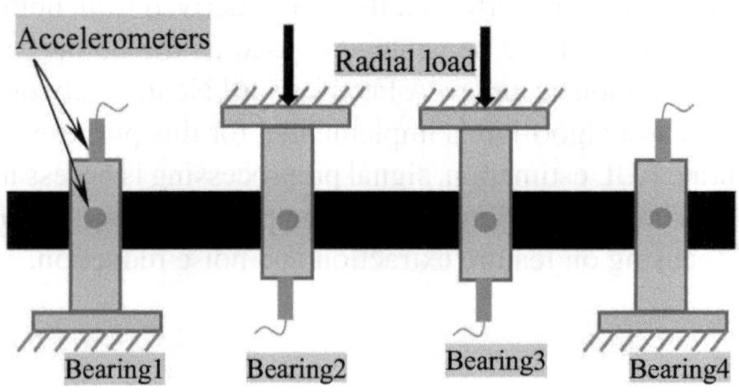

Fig. 1. Experimental system

Figure 2 presents the machine vibration signal recorded until a failure. In fact, it is a data acquisition representing the acceleration at a step level for 7 days every 10 min are equals to 1 data point. From this vibration signal, it is possible to take out the values or the characteristics that represent the degradation of the bearing.

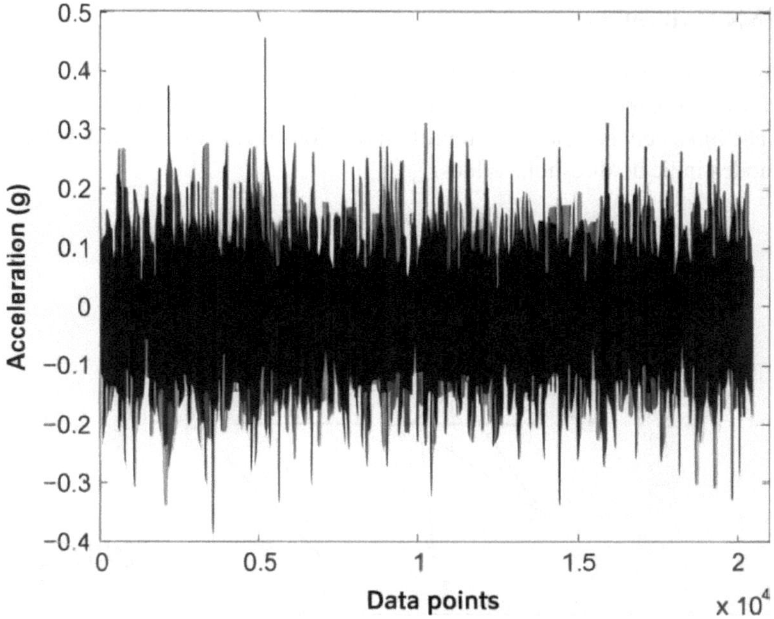

Fig. 2. Vibration signal recorded until failure

To detect a bearing failure, there are two values that are often used which are:

$$\text{RMS} = \sqrt{\frac{\sum_{f=1}^{f} Y(f)^2}{f}} \quad (1)$$

$$Kurtosis = \sqrt{\frac{\sum_{f-1}^{f} (Y(f) - Y_n)^4}{(f-1)Y_{std}^4}} \quad (2)$$

As defined *Y(f)* is a sequence of signals for $f = 1, 2, \ldots, f$, then f is the number of data points.

The typical deviation Y_{std} and mean value Y_n of the indicator are defined by:

$$\text{Y}_{\text{std}} = \sqrt{\frac{\sum_{f=1}^{f} (Y(f) - Y_n)^4}{(f-1)}} \quad (3)$$

$$\text{Y}_{\text{m}} = \frac{1}{K} \sum\nolimits_{f=1}^{f} Y(f) \quad (4)$$

3 Artificial Neurons Network

The training and verification of "ANN" is a very important step that we will develop as follows:

3.1 FFNN Verification

Multi-layer Feed-Forward Neural Networks (FFNN), commonly used in scientific research, often comprise two or three levels. Figure 3 presents the architecture of an FFNN. The links between elements represent relationships with assigned weights, while certain nodes indicate threshold values.

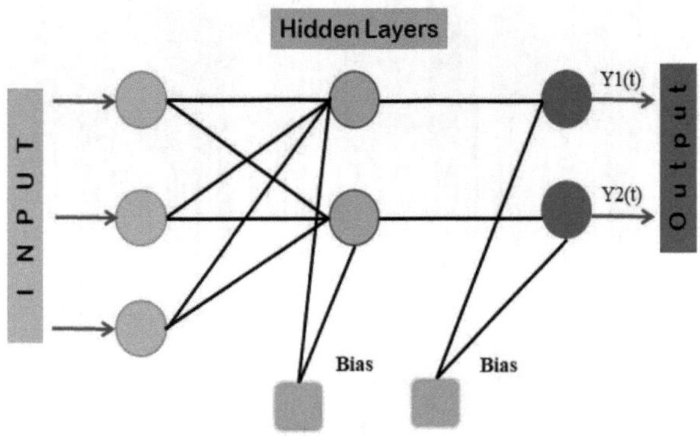

Fig. 3. Chart graph for "FFNN".

The FFNN is representing by this mathematical way:

$$Y_i = \varphi_o[C\varphi_h(B.U_i + b_h) + b_o] \tag{5}$$

All Components of a feed forward Neural Network presented by Fig. 4

3.2 Suggested ANN Prediction Approach

The Remaining Useful Life of a bearing is governed by a nonlinear dependency. Accurately predicting it requires a robust tool capable of identifying the relationship between input data from the bearing and it's Remaining Useful Life. Artificial neural networks are proposed for this purpose, as they are highly effective in modeling the nonlinear functions of complex systems.

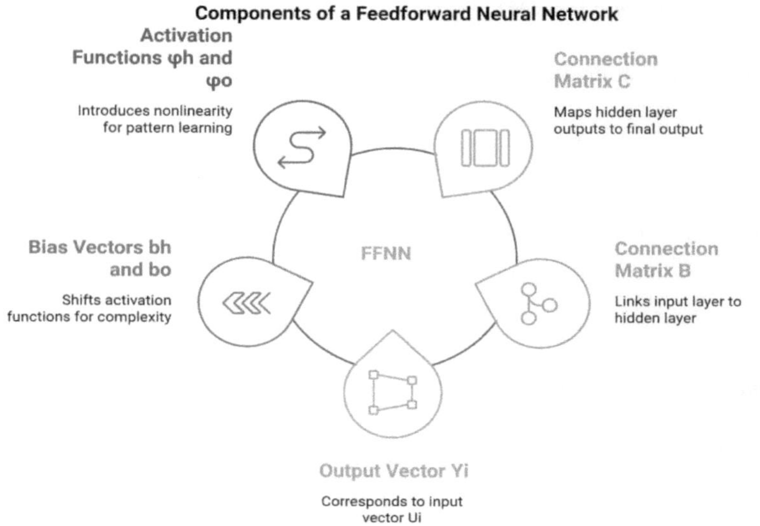

Fig. 4. Components of a feedforward Neural Network

The suggested "ANN" model employs fitted measurements for input data instead of actual measurements. This is because the measurement data obtained from practical applications often contains external noise, which affects the measurement data. Therefore, we use adjusted measurement data as input to represent bearing degradation characteristics. The fit measure uses the Weibull failure rate function, a robust tool for modeling machine reliability. Weibull's failure rate equation is:

$$G(t) = \frac{b(t)}{1 - B(t)} \tag{6}$$

B(t) represents the probability density function, while *b(t)* corresponds to the failure rate function. Both are functions of cumulative density, γ, and η.

$$B(t) = 1 - \exp\left[-\left(\frac{t}{\gamma}\right)^{n}\right] \text{t} \geq 0 \tag{7}$$

Table 1 illustrates the results from the Weibull parameter estimates for determining the values of *RMS* and *kurtosis*.

Table 1. Parameter estimation results

Weibull failure rate	Parameter estimated	
RMS	$\gamma = 0.1198$	$\eta = 1.9367$
Kutosis	$\gamma = 0.8947$	$\eta = 1.2008$

- ***Weibull failure rate for RMS:***

  - For the present inspection, we have:

$$Z_i^1(t) = \frac{f_i^1}{1 - F_i^1} \tag{8}$$

  - For the previous inspection, we have:

$$Z_{i-1}^1(t) = \frac{f_{i-1}^1}{1 - F_{i-1}^1} \tag{9}$$

- ***Weibull failure rate for kurtosis:***

  - For the current inspection, we have:

$$Z_{i-1}^2(t) = \frac{f_{i-1}^2}{1 - F_{i-1}^i} \tag{10}$$

  - For the previous inspection, we have:

$$Z_i^2(t) = \frac{f_i^2}{1 - F_i^i} \tag{11}$$

3.3 Training Phase of the ANN

Six inputs are integrated into the network for "ANN" training, as follows:

- t_i with t_{i-1}: Current and previous inspection times.
- Z_i^1 With Z_{i-1}^1 are the RMS values of measurements adjusted during the current and previous inspection.
- Z_i^2 With Z_{i-1}^2 is the kurtosis values of measurements adjusted during the current and previous inspection respectively.

The "ANN" uses the current and previous time values to estimate the Remaining Useful Life "RUL" of the bearing.

The output of the "ANN" is the normalized life percentage T_i, chosen as the output variable.

This metric directly reflects the bearing's health. A value of 100% indicates that the bearing has reached complete failure.

Figure 5 illustrates the architecture of the proposed "ANN".

The values of the "*RMS*" and "*kurtosis*" measurements taken for bearings B1, B2, B3 and B4 are shown in Fig. 6. At the conclusion of the test, a failure of the outer race occurs in bearing B1. For thus, we have chosen to study only the behavior of bearing B1.

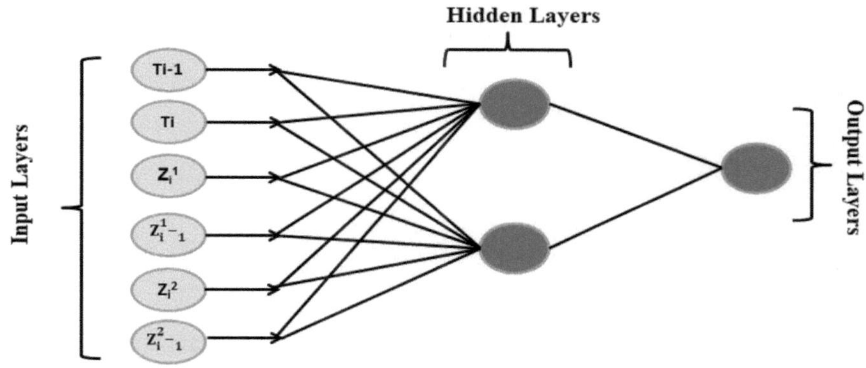

Fig. 5. Structure of "ANN"

Training a neural network is a critical data analysis process. It is essential to halt the training as soon as performance indicators exceed acceptable thresholds. Overfitting occurs when a model performs exceptionally well on the training data but shows poor performance on unseen data. To mitigate this issue, cross-validation is employed; this is a widely recognized practice in machine learning. Cross-validation involves using two distinct datasets: one for training the network and another for validating its performance. Overfitting is characterized by a specific behavior in the mean squared error "MSE" of the validation set. Initially, the MSE decreases as the model begins to learn patterns in the data. However, after reaching a minimum point, the "MSE" begins to rise, indicating that the model is no longer generalizing well to unseen data. In contrast, the "MSE" of the training set continues to decline throughout this process, suggesting that the model is fitting the training data too closely, capturing noise rather than underlying patterns. When the "MSE" of the validation set starts to increase, it signals that the regression algorithm is overfitting the training data. To prevent this, training should be halted at this point, ensuring that the model maintains its ability to generalize effectively to new, unseen data.

"MSE" is utilized to current network performance to define an optimal network. The equation is defined by:

$$MSE = \frac{1}{N}\sum\nolimits_{i}^{N} (e_i)^2 = \frac{1}{N}\sum\nolimits_{i=1}^{N} (t_i - a_i)^2 \tag{12}$$

where:

e_i: presents the error,

a_i: represents present value,

t_i: represents a target value,

N: presents an amount of data.

In this study, the data is divided: a training set and a validation set, to address the issue of over fitting. The training set consists the original dataset from the input, while the validation set is corrupted by 5% traffic. Additionally, the actual output of the "FFNN" is normalized to a range between 0 and 1 to maintain uniformity in variable sizes and prevent numerical instability.

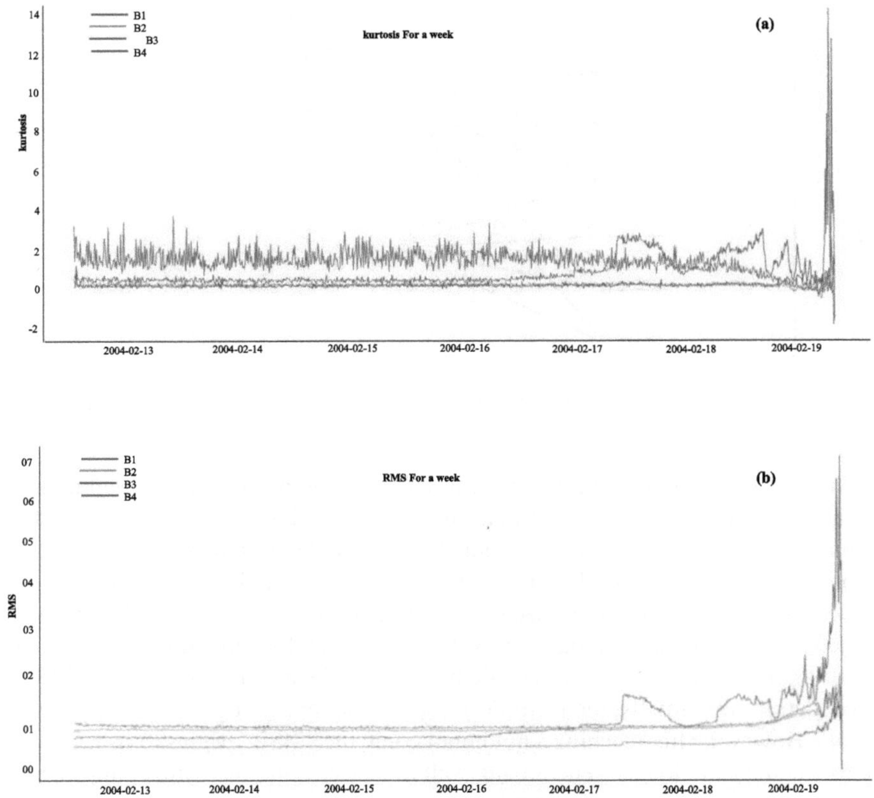

Fig. 6. Kurtosis (a), RMS (b) values for a week.

The artificial neural network is trained and verified to minimize the validation error. The learning and validation of "ANN" are configured from 2 to 20 hidden nodes. The network that produces the smallest validation error is chosen as the optimal network. The cause is that the network can still function properly even if the input in a particular stream is corrupted.

3.4 Configuration of ANN Model

The configuration of the Artificial Neural Network "ANN" employs the sigmoid transfer function in the hidden layer, while the output layer utilizes the linear Purelin transfer function. This combination enables the network to achieve robust performance. To optimize the training process, the Levenberg-Marquardt algorithm is used as the optimization technique.

This algorithm is existed to minimize functions which are nonlinear least squares suitable for estimating a nonlinear model. It can approximate the second-order convergence point without requiring the computation of the Hessian matrix. If the performance function attains a least squares suitable, the approximation of Hessian matrix can be written as:

$$H = J^T J \tag{13}$$

where J is a Jacobian matrix. The hessian matrix much more complex so it's better to use this algorithm which can be calculated using a standard back-propagation method.

The "Levenberg-Marquardt" algorithm employs an approximation of the Hessian matrix in the Newton-type update, as shown below:

$$X_{k+1} = X_k\{-[J^T J + \mu I]^{-1}\}J^T e \tag{14}$$

We define the gradient $g = J^T \cdot e$, where e, is a vector of network errors.

When a scalar $\mu = 0$ then it becomes similar to newton's method by using the approximate "hessian matrix".

3.5 Global Procedure

Figure 7 illustrates the overall methodology for the predictive model development. The process begins with the input of measurement values (Weibull, RMS, Kurtosis), followed by data association and the creation of a training dataset for the artificial neural network (ANN). The ANN undergoes training, validation, and model optimization to improve prediction accuracy. The final trained ANN is designed to predict a target percentage with a 5% accuracy threshold.

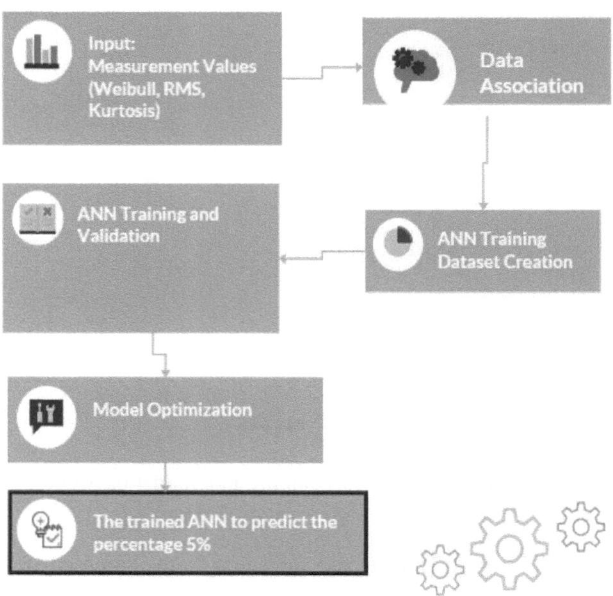

Fig. 7. Global procedure

4 Results

The proposed network has been trained and validated to assess its performance. The outcomes of the training and validation processes are presented in Table 2.

Figures 8 show the Training and validation performances. The ' + ' symbol represents the actual output, while the '•' symbol indicates the predicted output. The X-axis shows the bearing's operating time, and the Y-axis represents the bearing's life percentage

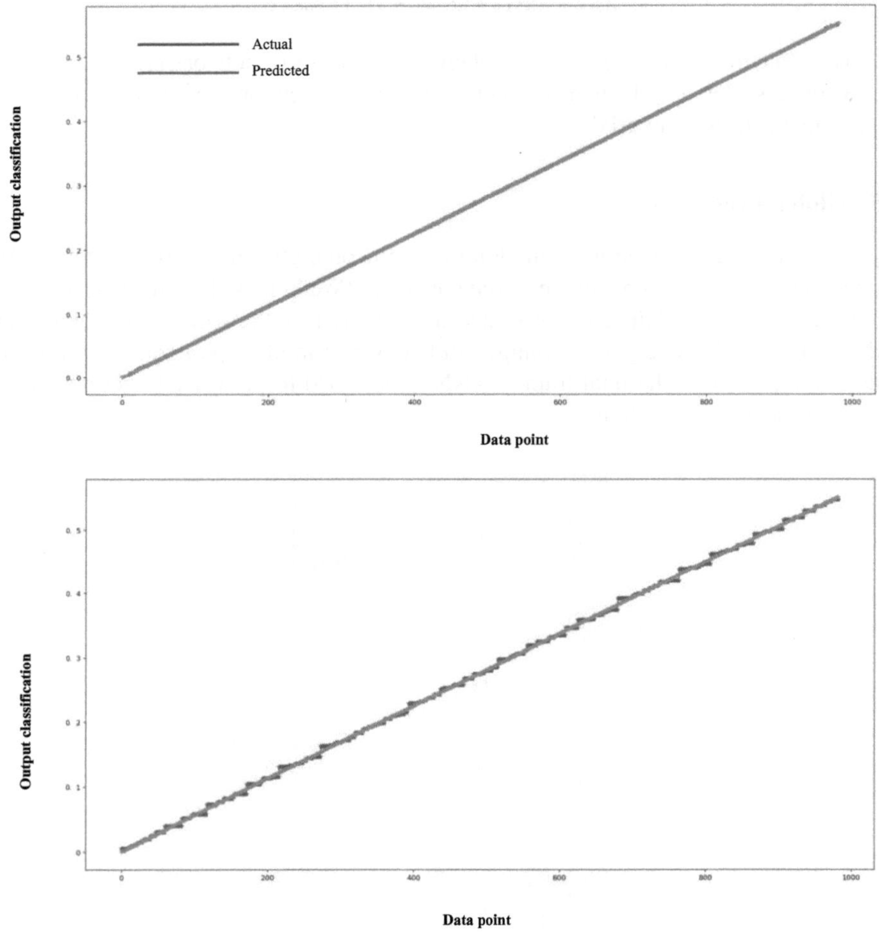

Fig. 8. Outcomes of the Training and Validation Model

Table 2. Errors in Training and Validation for the Proposed Model

Network model	Hidden node	Training error	Validation error
FFNN	*2*	*9.9 E−13*	*3.1 E−12*

in standardized form. For instance, after the bearing has operated for 6000 min, the remaining life is about 35%, implying the Remaining Useful Life "RUL" is roughly 65%.

Figure 9 illustrates the comparison between actual and predicted bearing degradation values using our Feedforward Neural Network (FFNN) model. The test was conducted on 100 unseen data points from the training set, with an added 10% noise disturbance to simulate real-world variations. The alignment between predicted and actual values demonstrates the robustness of the model in capturing degradation trends.

The close correlation between both datasets confirms the model's ability to generalize beyond training data, making accurate predictions even in the presence of noise. This predictive reliability is critical for real-world applications, as it enables early failure detection and supports proactive maintenance strategies.

From an operational perspective, the ability to forecast bearing degradation with high precision ensures timely interventions before reaching critical failure. This extends the bearing's functional lifespan, minimizes downtime, and reduces maintenance costs. The model's accuracy also reinforces the effectiveness of Artificial Neural Networks (ANNs) in Remaining Useful Life (RUL) estimation, providing a reliable tool for predictive maintenance.

Recent studies, such as the work of Fassi et al. (2023), highlight that incorporating physical knowledge into machine learning frameworks enhances the reliability and consistency of predictive models while reducing dependence on extensive training datasets. However, in our study, we opted for a pure ANN approach due to its ability to adapt efficiently to varying operating conditions and capture the complex dynamics of the

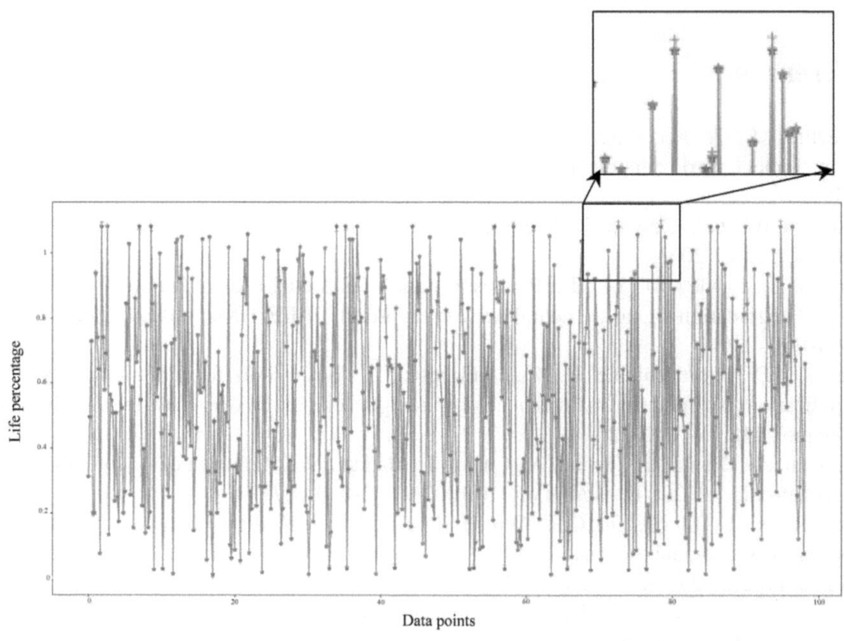

Fig. 9. Testing Process Evaluation (— Actual, — Predicted)

analyzed bearing system. This choice ensures a balance between computational efficiency and predictive accuracy, making it a suitable alternative to purely physics-based or hybrid models in our specific application.

Using the Artificial Neural Network "ANN" methodology, it is possible to accurately determine and predict the percentage of bearing failure throughout its operational lifespan. This predictive capability allows for proactive maintenance strategies, ensuring that potential failures can be addressed before they occur. Notably, a bearing is considered to have failed once it reaches 100% of its designated service life, highlighting the importance of continuous monitoring and forecasting to extend its functional longevity.

5 Conclusion

This study presents an innovative approach utilizing Artificial Neural Networks "ANN" to enhance the accuracy of bearing life predictions. The proposed model incorporates critical input variables, including time, Weibull failure rates, and the root mean square "RMS" and kurtosis values derived from both current and historical data points. By employing these specific inputs, the model effectively minimizes degradation noise, resulting in more reliable and precise prognosis of bearing lifespan. The output of the model is represented as a normalized percentage of remaining useful life "RUL".

The results from our experiments indicate that the "ANN" model not only demonstrates robust reliability in predicting the "RUL" of bearing failures but also significantly contributes to the optimization of maintenance strategies. This predictive capability allows for timely interventions, ultimately reducing downtime and enhancing operational efficiency in industrial applications.

References

Basseville, M., Cordier, M.O.: Surveillance et diagnostic de systèmes dynamiques : approche complémentaire du traitement de signal et de l'intelligence artificielle. Rapport INRIA N°2861 (1996)

Dubuisson, B., et al.: Diagnostic, Intelligence Artificielle et reconnaissance des formes, Editions Hermès, Paris (2001)

Gebraeel, N., Lawley, M., Liu, R., Parmeshwaran, V.: Residual life predictions from vibration-based degradation signals: a neural network approach. IEEE Trans. Industr. Electron. **51**(3), 694–700 (2004)

Popescu, T.D., Aiordachioaie, D., Culea-Florescu, A.: Basic tools for vibration analysis with applications to predictive maintenance of rotating machines: an overview. Int. J. Adv. Manuf. Technol. **118**, 1–17 (2022)

Hammami, A., Djemal, F., Hmida, A., Chaari, F., Haddar, M.: Maintenance 4.0 of wind turbine. Mechatronics (2020). https://doi.org/10.1007/978-3-030-46729-6_1

Huang, R., Xi, L., Li, X., Liu, C.R., Qiu, H., Lee, J.: Residual life predictions for ball bearings based on self-organizing map and back propagation neural network methods. Mech. Syst. Signal Process. **21**(1), 193–207 (2007)

Lee, J., Qiu, H., Yu, G., Lin, J.: Rexnord Technical Services, "Bearing Data Set", IMS, University of Cincinnati. NASA Ames Prognostics Data Repository, Moffett Field, CA (2020). http://ti.arc.nasa.gov/tech/dash/pcoe/prognostic-data-repository/

Li, C.J., Lee, H.: Gear fatigue crack prognosis using embedded model, gear dynamic model and fracture mechanics. Mech. Syst. Signal Process. **19**, 836–846 (2005)

Li, Y., Billington, S., Zhang, C., Kurfess, T.R., Danyluk, S., Liang, S.Y.: Adaptive prognostics for rolling element bearing condition. Mech. Syst. Signal Process. **13**, 103–113 (1999)

Li, Y., Kurfess, T.R., Liang, S.Y.: Stochastic prognostics for rolling element bearings. Mech. Syst. Signal Processi. **14**, 747–762 (2000)

Pan, Y., Chen, J., Li, X.: Bearing performance degradation assessment based on lifting wavelet packet decomposition and fuzzy c-means. Mech. Syst. Signal Process. **24**(2), 559–566 (2010)

Pham, H.T., Tran, V.T., Yang, B.S.: A hybrid of nonlinear autoregressive model with exogenous input and autoregressive moving average model for long-term machine state forecasting. Expert Syst. Appl. **37**(4), 3310–3317 (2010)

Qiu, H., Lee, J., Lin, J., Yu, G.: Wavelet filter-based weak signature detection method and its application on rolling element bearing prognostics. J. Sound Vib. **289**(4–5), 1066–1090 (2006)

Satish, B., Sarma, N.D.R.: A fuzzy BP approach for diagnosis and prognosis of bearing faults in induction motors. IEEE Power Eng. Soc. General Meet. **3**, 2291–2294 (2005)

Si, X.S., Wang, W., Chen, M.Y., Hu, C.H., Zhou, D.H.: A degradation path-dependent approach for remaining useful life estimation with an exact and closed-form solution. Eur. J. Oper. Res. **226**(1), 53–66 (2013)

Tian, Z.: An artificial neural network method for remaining useful life prediction of equipment subject to condition monitoring. J. Intell. Manuf. (2009)

William, P.E., Hoffman, M.W.: Identification of bearing faults using time domain zero-crossings. Mech. Syst. Signal Process. **25**(8), 3078 (2011)

Fordal, J.M., Schjølberg, P., Helgetun, H., Skjermo, T.Ø., Wang, Y., Wang, C.: Application of sensor data based predictive maintenance and artificial neural networks to enable Industry 4.0. Adv. Manuf. **11**(2), 248–263 (2023)

Khalil, A.F., Rostam, S.: Machine learning-based predictive maintenance for fault detection in rotating machinery: a case study. Eng. Technol. Appl. Sci. Res. **14**(2), 13181–13189 (2024)

Fassi, Y., Heiries, J., Boutet, J.: Toward physics-informed machine-learning-based predictive maintenance for power converters—a review. IEEE Transactions on (2023)

Impact of Design Parameters on the Dynamic Behavior of Composite Shafts

Salwa Benali[1,2](✉), Taissir Hentati[1], and Slim Bouaziz[1]

[1] Laboratory of Mechanics, Modeling and Production (LA2MP), National Engineering School of Sfax, University of Sfax, Sfax, Tunisia
salwa.benali96@yahoo.com

[2] National Engineering School of Gabes, University of Gabes, Gabés, Tunisia

Abstract. Modal analysis critically enables engineers to characterize composite shafts' vibrational dynamics and optimize their mechanical performance by harmonizing fiber orientation, stacking sequence, and stiffness, ensuring robust rotating system design. This paper investigates the modal characteristics of composite shafts, highlighting the impact of design parameters on natural frequencies values.

Using Hamilton's principle, the governing equations of motion for the shaft were formulated, integrating stiffness and damping properties. A finite element model based on the Equivalent Single Layer Theory (ESLT) was implemented to model the composite shaft's anisotropic properties.

This research emphasizes the influence of several factors such as material types, fiber orientation and stacking sequence, on the modal behavior of composite shafts. Various composite materials, including Bore/Epoxy and Carbon/Epoxy, were examined to determine their natural frequencies.

Results indicate that optimizing key parameters can significantly enhance the composite shaft's performance. Specifically, the natural frequencies of the Carbon/Epoxy (C/E) shaft reach their maximum when fibers are aligned at 0°, while adding additional layers increases stiffness, raising natural frequencies further. It is significant that higher material stiffness directly correlates with elevated natural frequencies. These findings collectively demonstrate that strategic parameter selection improves stiffness-to-weight ratios, refines vibrational response, and mitigates resonance risks in rotating systems.

Keywords: FEM · Modal analysis · Composite · Fiber orientation · material type · stacking sequence

1 Introduction

Drive shafts are essential elements in aerospace and automotive applications, responsible for transmitting torque from the engine to the propellers or wheels. Traditional drive shafts have been manufactured from metal alloys. However, the growing demand for more efficient machines has led to the development of composite shafts as an alternative. These latter offer several advantages, such as lower weight, higher stiffness, and

A. Hammami et al. (Eds.): MOSCOSSEE 2024, ACM 23, pp. 168–177, 2025.
https://doi.org/10.1007/978-3-032-00930-2_16

better corrosion resistance. However, the use of composite shaft introduces new challenges as understanding their modal behavior. In this context, several research have been performed to investigate the modal behavior of composite shafts using both numerical and experimental techniques.

The study conducted by (Fabian et al., 2022),focused on investigating the modal parameters of two shafts, the first one made of Carbone/epoxy composite and the other one was made of structural steel, using an Operational Modal Analysis (OMA) technique. The results demonstrate the effectiveness of the OMA technique in characterizing the dynamic behavior of composite shafts. Another study developed by (Filipeson et al., 2024) focused on determining the experimental and theoretical natural frequencies and mode shapes of a composite drive shaft of a Toyota Qualis. The composite shaft is manufactured from an aluminum matrix, reinforced with vanadium and tungsten fibers. Results showed that the mass density of matrix and fibers, have significant effects on natural frequencies of the shaft. Moreover,(Rakesh Raushan, 2024) presents in their work the design of a hollow composite shaft for cars. The study reveals that the CF/PA66 shaft exhibits comparable natural frequencies and reduces vibrations compared to traditional steel drive shafts furthermore, (Bhirud Pankaj Prakash, 2014) analyzed the dynamic properties of an automotive steel and composite drive shaft. The composite shaft is formed by Kevlar. The researchers conducted finite element modal analysis in order to identify natural frequencies and shape mode of the composite drive shaft. Results showed that composite shaft natural frequencies are higher than that of steel material. A study given by (Ameen & Ali, 2020) involves analyzing the modal behavior of a circular composite shaft with free boundary conditions. To extract natural frequencies and mode shapes, the Experimental Modal Analysis (EMA) method is employed. A modal analysis conducted by (Xiu & Chang, 2024) focuses on enhancing the dynamic behavior of a tripod-ball type universal coupling composite shaft through with ANSYS Workbench. Results indicate that, the intermediate shaft, bell housing, and trident sleeve are the areas where the greatest deformation occurs. Another study conducted by (Wang, 2024) using a 3D FEM based on Timoshenko beam theory and the Lagrangian method. Authors studied a four-disk hollow shaft system. Results show that suitable damping has a consistent effect on the system during the start-up.

The present article aims to build upon the modal analysis of a composite rotating shaft formed by epoxy matrix reinforced with carbon fibers. We will present a detailed investigation of the influence of several parameters such as the stacking sequence, the fiber orientation and the number of layers on the natural frequencies of the shaft. The results will be compared with those obtained in the literature.

2 Mathematical Formulation

In order to study modal behavior of a composite rotating shaft, FEM is used. The shaft is discretized into 22 elements as shown in Fig. 1. Each element is composed of two nodes, each one being composed of four degrees of freedom.

Fig. 1. Finite element model

The variational formulation used in our FEM is extensively described in (Benali et al. 2023). Using the Hamilton principle, the motion equation of the shaft is expressed as:

$$[M]\{\ddot{q}\} + [C]\{\dot{q}\} + [K]\{q\} = \{F_{ext}\} \tag{1}$$

where [M], [K], and [C] represent, respectively, the mass, stiffness, and damping matrices given respectively by Eqs. 2, 3 and 4.

$$[M] = \begin{bmatrix} [M_e^1] & 0 & 0 & 0 & 0 & 0 \\ & [M_e^2] & 0 & 0 & 0 & 0 \\ & & [M_e^3] & 0 & 0 & 0 \\ & & & \ddots & 0 & 0 \\ & & & & [M_e^{21}] & 0 \\ Sym. & & & & & [M_e^{22}] \end{bmatrix} \tag{2}$$

$$[K] = \begin{bmatrix} [K_e^1] & 0 & 0 & 0 & 0 & 0 \\ & [K_e^2] & 0 & 0 & 0 & 0 \\ & & [K_e^3] & 0 & 0 & 0 \\ & & & \ddots & 0 & 0 \\ & & & & [K_e^{21}] & 0 \\ Sym. & & & & & [K_e^{22}] \end{bmatrix} \tag{3}$$

where $[M_e]$ and $[K_e]$ represents respectively the elementary mass and damping matrices given by (Benali et al., 2023)

$$[C] = \alpha[M] + \beta[K] \tag{4}$$

where α and β are real positive constants.

To determine the natural frequencies of the studied shaft, the equation of motion given in (1) was converted to a first-order form as (Benali et al., 2023).

$$[A]\{\dot{x}\} + [B]\{x\} = \{0\} \tag{5}$$

Matrices [A] and [B] are given by:

$$[A] = \begin{bmatrix} [M] & [0] \\ [0] & [M] \end{bmatrix} \tag{6}$$

$$[B] = \begin{bmatrix} [0] & [-M] \\ [K] & [C] \end{bmatrix} \tag{7}$$

The generalized equation for resolving the shaft eigenvalues λ, is mentioned by [Singh and Gupta 1996b],

$$\det([B] - \lambda[A]) = 0 \tag{8}$$

3 Results and Discussions

In order to compare the impacts of several parameters on composite shaft eigenfrequencies, a parametric study was performed. This analysis examined the specific influence of each factor, including: the stacking sequence of the composite layers, the number of layers, the type of composite material used and the stiffness of the bearings.

A hollow shaft made up of several layers formed by epoxy matrix reinforced by Carbon fibers has been studied. The mechanical characteristics of the composite material used are given in Table 1. Each layer has 1 mm as thickness. The shaft have 1m as length and 0.048 m as diameter as shown in Fig. 2.

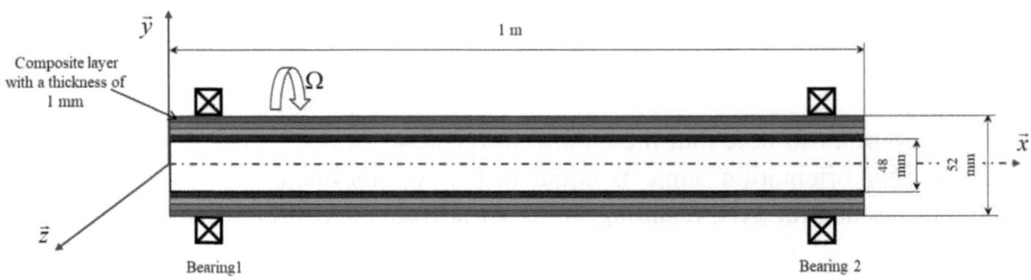

Fig. 2. Studied shaft (Benali et al., 2023)

Table 1. Properties of the composite material (Benali et al., 2024a, b)

Material	E_1 (GPa)	E_2 (GPa)	G_{ij} (GPa)	ν_{12}	ρ (kg/m^3)
Carbone/Epoxy	130	10	7	0.25	1500

3.1 Impact of Fiber Orientation Angle

To highlight the influence of modifying the fiber orientation angle on the natural frequencies of the shaft, we consider the same structure described in Fig. 2. The shaft consisted of eight symmetric (8S) Carbon/Epoxy (C/A) layers where the fiber orientation was modified: seven case studies are deduced. The eigenfrequency of each case study is computed and recapitulated in Table 2.

Also, we represent the variation of the first and second natural frequencies and the equivalent modulus functioning of the fiber orientation given by Fig. 3.

The expression of the equivalent modulus is formulated as a function of the fiber orientation angle α, (Benali et al., 2023) as presented in Eq. (9).

$$\overline{Q}_{11} = Q_{11}\cos^4\alpha + Q_{22}\sin^4\alpha + (2Q_{12} + 4Q_{66})\cos^2\alpha\sin^2\alpha \tag{9}$$

Table 2. Influence of fiber orientation on natural frequencies

Fiber orientation	F1(Hz)	F2(Hz)	F3(Hz)
[0] 8S	81	276	940
[±15] 8S	60.3	248.5	1024
[±30] 8S	46.5	192	792
[±45] 8S	26.4	133	670
[±60] 8S	22.6	93.4	385.9
[±75] 8S	20.5	82	327
[±90] 8S	19.3	64	212

It is important to note that the natural frequencies of the C/E shaft are maximized once the fiber orientation angle is equal to 0°. At this angle, the fibers are aligned parallel to the neutral axis, resulting in higher stiffness and strength characteristics. In contrast, they reach their minimum values in the case that this angle is equal to 90°. The variation in natural frequencies is reciprocally related to the change in the fiber orientation angle. The fibers are oriented perpendicular to the neutral axis, which effectively reduces the stiffness of the composite material. This reduction in stiffness leads to decreasing natural frequencies. This can be justified by the fact that the equivalent stiffness modulus decreases as the fiber orientation angle increases, as shown in Fig. 3.

3.2 Impact of Number of Layers

In this section, we aim to present the C/E shaft layers number effect on the natural frequencies' values. We study the structure illustrated in Fig. 2. The shaft initially made up of four layers, afterward eight layers, and lastly twelve layers.

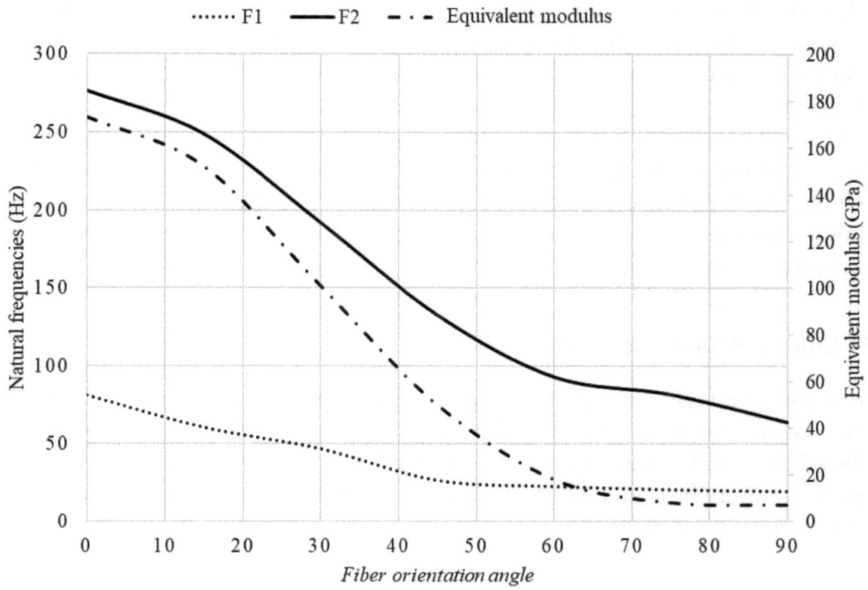

Fig. 3. Variation of natural frequencies and equivalent stiffness in composite materials as a function of fiber orientation angle

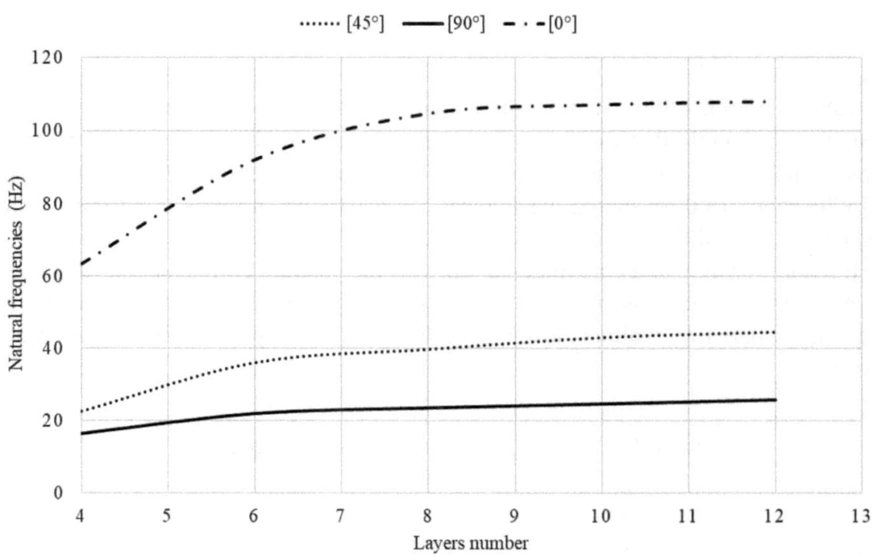

Fig. 4. Variation of the first natural frequency as a function of the number of layers

It is significant that adding additional layers leads to an increase in the stiffness of the shaft, resulting in higher natural frequencies as illustrated by Fig. 4. This can be explained by the combined effect of the geometric arrangement and material properties of the layers.

It can be seen from Fig. 4 that adding layers from 4 to 7 significantly enhances stiffness and natural frequencies. The increase of frequencies becomes less intensity when the number of layers exceeds 7. This indicates that the structural advantages of additional layers in the composite shaft may tend to stabilize, leading to slight increases in natural frequency.

3.3 Impact of Shaft Material

In this part, we highlight the impact of the shaft material on its natural frequencies of the system. A composite shaft comprised of various materials: Boron/Epoxy (B/E), Carbon/Epoxy (C/E), Glass/Epoxy (G/E), and Aluminum (Al) is studied where their properties are outlined in Table 3. The shaft is formed by four layers, classified in a stacking sequence of [0, 45, 45, 45].

Table 3. Materials properties (Benali et al., 2024a, b)

Material	E1 (GPa)	E2 (GPa)	Gij (GPa)	ν12	ρ (kg/m3)
B/E	211	24.1	6.9	0.36	1967
V/E	40	5	4	0.28	1800
Al	66	—	—	0.33	2680

Table 4 presents the first trio natural frequencies of the composite shaft. It can be seen from this table that the frequencies are affected by the categories of materials implemented. It is important to note that, the natural frequencies are similar for B/E and C/E owing to their shared mechanical characteristics in terms of stiffness. These frequencies exceed those achieved with G/E.

It's noteworthy that as the material stiffness increases, a corresponding increase in the natural frequencies is observed. This explains the observed hierarchy of natural frequency, where B/E exhibits the highest values, followed by C/E, while G/E shows the lowest.

Table 4. Effects of material type on natural frequencies

Material	F1 (Hz)	F2 (Hz)	F3 (Hz)
B/E	329	1122	2283
C/E	320	1115	2270
V/E	135	446	813

Now, we will examine the hybrid configurations: the shaft is formed by four composite layers and one layer of aluminum. The shaft stacking sequence is [0, 45, 45, 45]. The various cases examined are presented as follows:

- Case 1: An outer layer of aluminum and four layers of B/E,
- Case 2: An outer layer of aluminum and four layers of C/E,
- Case 3: An outer layer of aluminum and four layers of G/E,
- Case 4: An inner layer of aluminum and four layers of B/E,
- Case 5: An inner layer of aluminum and four layers of C/E,
- Case 6: An inner layer of aluminum and four layers of G/E.

The first, second and third natural frequencies of the shaft are summarized in Table 5.

Table 5. First three natural frequencies of hybrid shaft

Case study	F_1 (Hz)	F_2 (Hz)	F_3 (Hz)
Case 1	306	1031	1062
Case 2	297	1024	1046
Case 3	171	586	947
Case 4	293	1022	1053
Case 5	289	1016	1031
Case 6	176	591	952

As shown in Table 5, coating the B/E or C/E shaft with aluminum reduces the natural frequencies of the system. Conversely, the combination of aluminum and G/E results an increase in the natural frequencies of the shaft. Indeed, aluminum exhibits a higher specific density than the other materials used hence the addition of aluminum has the potential to change the distribution of mass and stiffness throughout the shaft, influencing its natural frequencies.

3.4 Impact of Bearing Elasticity

A shaft composed of four layers of C/E with a stacking sequence of [0, 45, 45, 45]. Two elastic bearings are considered which the radial stiffness are $K_{yy} = K_{zz} = K_p$, while the torsional stiffness are, $K_{yz} = K_{zy} = 0$.

Figure 5 presents the variation of the first natural frequency of the composite shaft as a function of bearings stiffness. It is noteworthy that the natural frequencies of the rotor rise as stiffness value of the bearing increases. Very high rigidity can lead to natural frequencies like a rotor supported by rigid bearings.

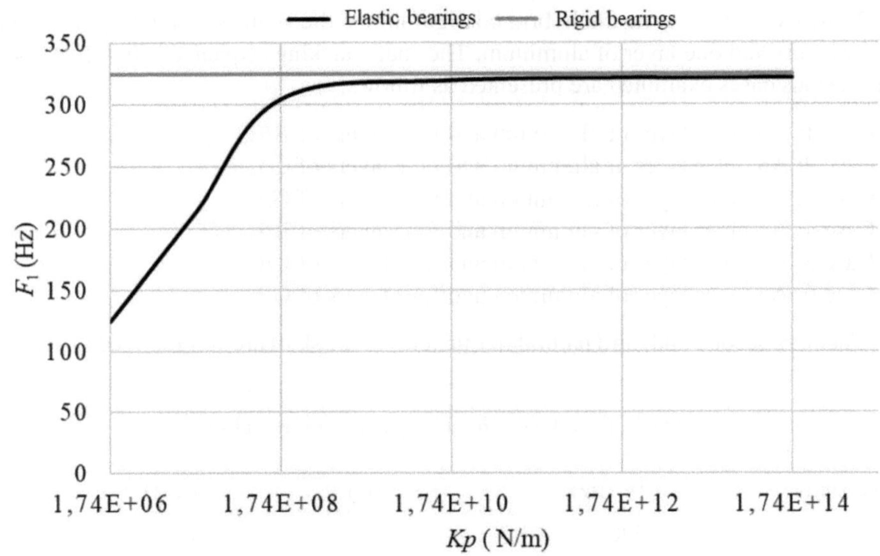

Fig. 5. First frequency variation as a function of bearing stiffness

4 Conclusion

In this work, a modal analysis of a composite shaft has been developed using the FEM. A parametric analysis is performed to examine the influence of several parameters: material type, fiber orientations, stacking sequence, number of layers, and bearing stiffness, on the natural frequencies of the system.

Results indicate that a refined selection of these parameters leads to improved system natural frequencies. Specifically, increasing the bearings stiffness, material rigidity characteristics, and number of layers contributes to increasing the shaft's natural frequencies.

References

Ameen, Y.M., Ali, J.K.: Theoretical and experimental modal analysis of circular cross-section shaft. IOP Conf. Ser. Mater. Sci. Eng. **745**(1) (2020)

Benali, S., Ben, A., Taissir, J., Slim, H., & Mohamed, B.: Stability analysis and enhancement of five-degree-of-freedom composite spindle with active magnetic bearing support. Int. J. Interactive Desi. Manuf. (IJIDeM) (2024)

Benali, S., Hentati, T., Ouaziz, S.B., Haddar, M.: Performance and efficiency of composite shafts supported by active magnetic bearings. J. Mech. Mater. Struct. **18**(5), 635–654 (2023)

Benali, S., Jdidia, A. Ben, Hentati, T., Bouaziz, S., Haddar, M.: Dynamic analysis and energy consumption modeling of a composite spindle with active magnetic bearings for a sustainable manufacturing. Proc. Inst. Mech. Eng. Part B J. Eng. Manuf. **6** (2024)

Prakash, B.P.: Analysis of drive shaft. Int. J. Mech. Prod. Eng. **2**(2) (2014)

Fabian, N.-A., Guillermo, S.-A.H., Diego, V.-B.: Evaluación de los parámetros modales de dos ejes, uno fabricado con materiales compuestos y otro en acero estructural, mediante la implementación de una técnica de análisis modal operacional (OMA). J. Eng. Sci. **25**, 83–92 (2022)

Filipeson, A.D., Xavier, S.B., Anyika, T.A.: Stress concentration and vibration modal analysis of propeller shaft using finite element analysis. Int. J. Appl. Eng. Technol. **6**, 59 (2024)

Raushan, R., Dhande, K.K., Jamadar, N.I.: Modal analysis of carbon fiber reinforced polyamide66 drive shaft using analytical and finite element approach. Int. J. Interact. Des. Manuf. (2024). https://doi.org/10.1007/s12008-024-01854-7

Wang, Y.: Model verification and vibration analysis of the four-disk hollow flexible shaft rotor system. Int. J. Mech. Sci. **268**, 109051 (2024)

Xiu, X., Chang, D.: Modal analysis of tripod-ball type universal coupling composite transmission shaft system. J. Phys. Conf. Ser. **2787**(1), 012001 (2024)

Integration of Artificial Intelligence: An Innovative Synergy for Vibration Reduction of Car Suspension System

Fatma Hmida[1,3](✉), Hassen Trabelsi[1,2], Fathi Djemal[1,3], and Rawnak Omar[1,3]

[1] Laboratory of Mechanics, Modelling and Production (LA2MP), National School of Engineering of Sfax, University of Sfax, Sfax, Tunisia
fhmida210@gmail.com

[2] Higher Institute of Industrial Systems of Gabès (ISSIG), University of Gabes, Gabes, Tunisia

[3] Laboratory of Mechanics, Modelling and Production (LA2MP), National School of Engineering of Gabes, University of Gabes, Gabes, Tunisia

Abstract. This study explores the innovative integration of artificial intelligence (AI) into the preliminary design of automotive suspension systems, focusing on vibration reduction and performance optimization. Unlike traditional methods, which often rely on static models and costly iterative testing, this research proposes a dynamic and data-driven approach, leveraging AI capabilities to enhance the precision and efficiency of the design process. A comprehensive parametric study is conducted to identify and analyze the most influential suspension parameters, such as stiffness, damping, geometry, and mass distribution. The Anaconda platform is used for data processing, modeling, and simulation, enabling a systematic analysis of the interactions between these parameters and their impact on the system's dynamic behavior. AI algorithms facilitate predictive analysis of vibrational behaviors and real-time adjustments, optimizing the system's response to varying road conditions. This approach not only enhances ride comfort but also contributes to the durability of components. Furthermore, by narrowing the intervals of each parameter through AI integration, the study ensures more precise optimization compared to traditional parametric studies. Detailed case studies demonstrate how AI outperforms conventional techniques in terms of precision, adaptability, and efficiency. Ultimately, the combination of parametric analysis and AI technologies offers new insights into the design of smarter, more efficient suspension systems, paving the way for more sustainable and innovative solutions in automotive engineering.

Keywords: car suspension · parametric study · dynamic · Machine Learning

1 Introduction

Suspension systems are fundamental to the performance and safety of modern vehicles, directly influencing both driving comfort and stability. By effectively connecting the wheels to the chassis, these systems absorb shocks from uneven surfaces, ensuring a

A. Hammami et al. (Eds.): MOSCOSSEE 2024, ACM 23, pp. 178–189, 2025.
https://doi.org/10.1007/978-3-032-00930-2_17

smooth ride while maintaining optimal tire contact with the road (Seifi et al., 2022). The efficiency of a suspension system is a critical determinant of passenger comfort and vehicle dynamics, underscoring its importance in automotive design (Alexandru and Alexandru, 2011).

In the traditional design process for suspension systems, the steps often follow a sequential, iterative cycle of "modeling-simulation-verification-validation." This cycle enables the validation of preliminary concepts and exploration of potential solutions, yet it has significant limitations. Sizing phases, often divided into static and dynamic analyses, are addressed separately, which can lead to over-dimensioning and suboptimal design. Without real-time adjustments, optimizing dimensional choices becomes labor-intensive and less precise (Siala et al., 2020; Mortazavi & Togan, 2017).

The rapid progress in machine learning (ML) techniques has greatly expanded their application across various engineering fields, especially in optimization and control systems. A Type-2 fuzzy neural network controller was developed to reduce vibrations caused by rough terrain on nonlinear vehicle suspensions (Taghavifar et al., 2019). A deep belief network (DBN) information fusion method was created for diagnosing thruster faults in deep-sea vehicles, addressing variable fault patterns (Zhu et al., 2021). An adaptive radial basis function neural network (RBFNN) was employed for structural reliability analysis, showcasing the versatility of ML (Jing et al., 2019). Particle swarm optimization (PSO) and sequential quadratic programming were utilized to optimize vehicle suspension parameters (Drehmer et al., 2015). An Elman neural network combined with PSO was applied for short-term power load forecasting (Xie et al., 2020).

This chapter aims to delve into the significance of parametric studies on suspension systems, highlighting the most influential parameters that shape their performance. We will explore the transformative role of data analytics in optimizing vehicle comfort and stability. By systematically exploring these key factors, we aim to generate valuable insights that can inform enhanced design and optimization strategies in the field of suspension system engineering.

2 Quarter Suspension System

The car suspension system employs fixed mechanical components to mitigate shocks. While this solution is straightforward and cost-effective, with recognized reliability, its inability to adapt to varying road conditions limits both comfort and driving performance. This work analyzes a simplified model of a car's suspension system to understand how it handles bumps and uneven roads. This simplified model, called a quarter-car model, focuses on the vertical movement of one wheel and its connected part of the car body. As shown in Fig. 1, the model considers two main parts: the sprung mass m_c (the car body) and the unsprung mass m_h (the wheel and axle). These two parts are connected by a spring k_b and a damper c_p. Additionally, the contact between the wheel and the road surface is modeled by a linear stiffness k_w which work together to absorb shocks. The road surface, which is constantly changing, acts as a force on the wheel, causing the entire system to move up and down (Barethiye, 2024).

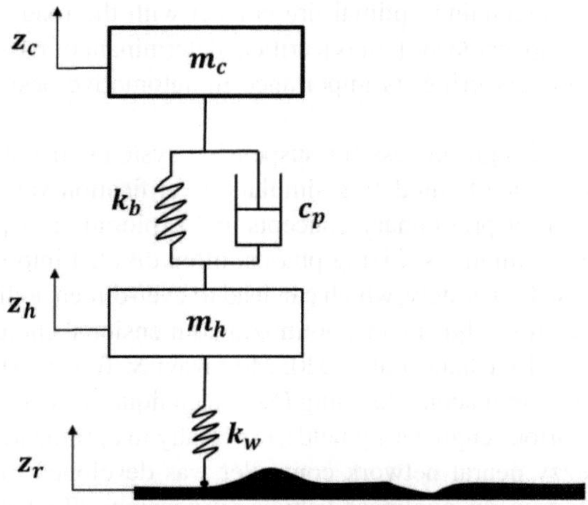

Fig. 1. Dynamic model of suspension system

2.1 Equations of Motion

The equations of motion are derived using Lagrange's method. In this context, let T, V, and D represent the kinetic energy, potential energy, and damping energy of the system, respectively. Therefore, they can be expressed as follows:

$$T = \frac{1}{2} m_h \dot{z}_h + \frac{1}{2} m_c \dot{z}_c \tag{1}$$

$$U = \frac{1}{2} k_b (z_h - z_c)^2 + \frac{1}{2} k_w (z_h - z_r)^2 \tag{2}$$

$$D = \frac{1}{2} c_p (\dot{z}_c - \dot{z}_h)^2 \tag{3}$$

The equations of motion are represented by the following equations:

$$m_c \ddot{z}_c + c_p (\dot{z}_c - \dot{z}_h) + k_b (z_c - z_h) = 0 \tag{4}$$

$$m_h \ddot{z}_h - c_p (\dot{z}_c + \dot{z}_h) + k_b (z_h - z_c) + k_w (z_h - z_r) = 0 \tag{5}$$

The equations of motion can now be expressed in matrix form as a function of the state vector:

$$M\ddot{z} + C\dot{z} + kz = F \tag{6}$$

With the computed matrices:

$$\begin{bmatrix} m_c & 0 \\ 0 & m_h \end{bmatrix} + \begin{bmatrix} c_p & -c_p \\ -c_p & c_p \end{bmatrix} + \begin{bmatrix} k_b + k_w & -k_b \\ -k_b & k_b \end{bmatrix} = \begin{bmatrix} 0 \\ 0 \end{bmatrix} \tag{7}$$

Equation (6) can be rewritten in the state space form as:

$$\dot{z} = Az + Wz_r(t)\ z = \begin{bmatrix} z_c \\ \dot{z}_c \\ z_h \\ \dot{z}_h \end{bmatrix} \tag{8}$$

The matrix A for the state equation system is:

$$A = \begin{bmatrix} 0 & 1 & 0 & 0 \\ \frac{-k_b}{m_c} & \frac{-c_p}{m_c} & \frac{k_b}{m_c} & \frac{c_p}{m_c} \\ 0 & 0 & 0 & 1 \\ \frac{k_b}{m_h} & \frac{c_p}{m_h} & \frac{-(k_b+k_w)}{m_h} & \frac{-c_p}{m_h} \end{bmatrix} \tag{9}$$

The vector of external forces W (which includes the disturbance) is given by:

$$w = \begin{bmatrix} 0 \\ 0 \\ 0 \\ \frac{k_w}{m_h} \end{bmatrix} \tag{10}$$

2.2 Shape Road

In this work, road disturbances are represented as external excitations $z_r(t)$. To simulate specific irregularities, the model includes a gentle 0,2 m wave lasting 2 s to represent mild bumps, followed by a sudden 0,1 m drop at 6 s to mimic a sharp pothole (Fig. 2).

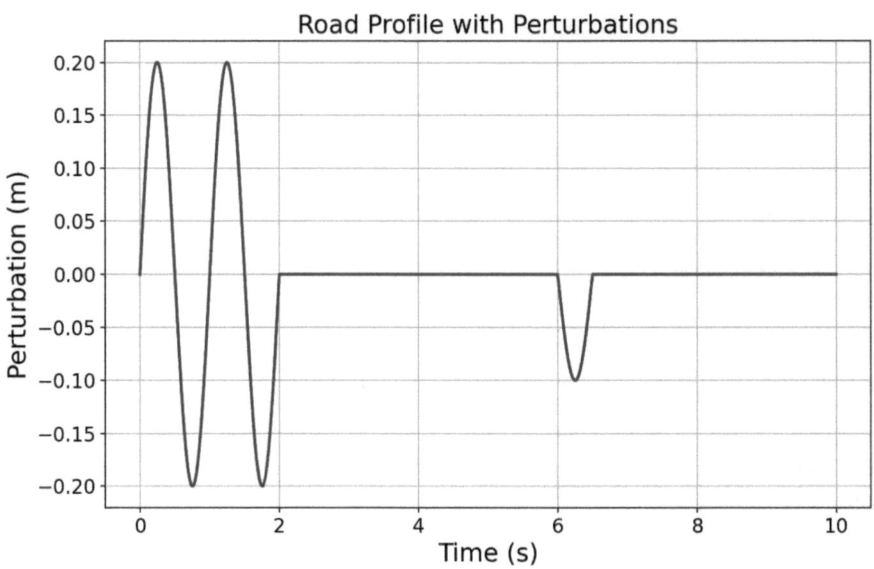

Fig. 2. Shape of the road disturbance

3 Dynamic Simulation of Suspension System

This study focuses on the simulation of suspension system responses to road disturbances. The values presented in Table 1 were used to guarantee the accuracy of the results. Accurately illustrating the dynamic behavior of the suspension system required a rigorous approach that integrated all relevant equations into a Python model. We obtain the reference curves shown in Fig. 3.

Table 1. Parameters values (Issa and Samn, 2022)

Variables	Description	Value	units
k_b	Spring stiffness	19960	N/m
m_c	Vehicle mass	288,9	kg
k_w	Tire stiffness	155900	N/m
m_h	Wheel mass	28,58	kg
c_p	Viscous damping coefficient	1861	N·s/m

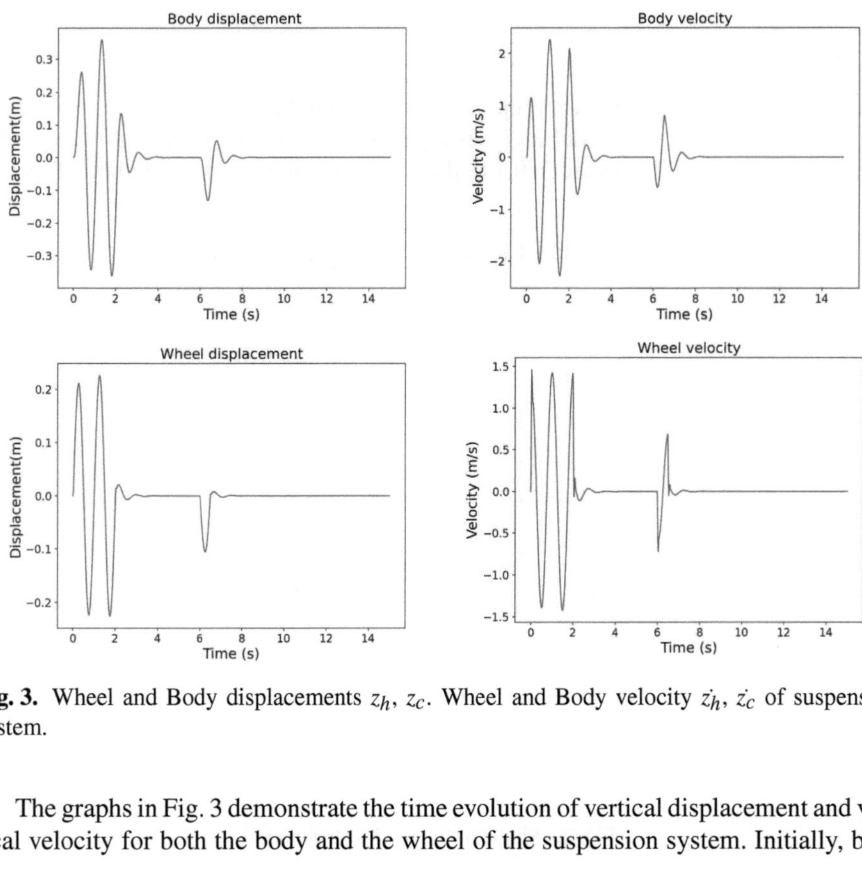

Fig. 3. Wheel and Body displacements z_h, z_c. Wheel and Body velocity $\dot{z_h}$, $\dot{z_c}$ of suspension system.

The graphs in Fig. 3 demonstrate the time evolution of vertical displacement and vertical velocity for both the body and the wheel of the suspension system. Initially, both components experience significant vertical displacement, reflecting the impact of external forces or disturbances. As the simulation progresses, the amplitude of the oscillations gradually decreases, indicating the system's stabilization. For the body, the maximum vertical displacement reaches 0,360 m, while the maximum velocity is approximately 2,2 m/s. These values represent the peak dynamic response before the system begins to stabilize. Similarly, the wheel's displacement reaches 0,226 m at its peak, with the maximum velocity around 1,4 m/s. The stabilization process leads to convergence, with the body reaching a steady state at 7,84 s and the wheel at 8,16 s. This signifies that both components of the system have reached their equilibrium positions, and the dynamic behavior is now governed by the system's natural frequencies and damping properties.

4 Methodologies

In this study, a comprehensive data analysis is conducted to support a detailed parametric investigation of the suspension system. The influence of various parameters on the system's overall performance is meticulously examined through this rigorous methodology. By identifying the complex relationships between these parameters and the observed outcomes, precise recommendations for system optimization are formulated. This systematic approach, illustrated by the process diagram in Fig. 4, ensures a deep understanding of the underlying dynamics and facilitates data-driven decision-making based on robust analyses.

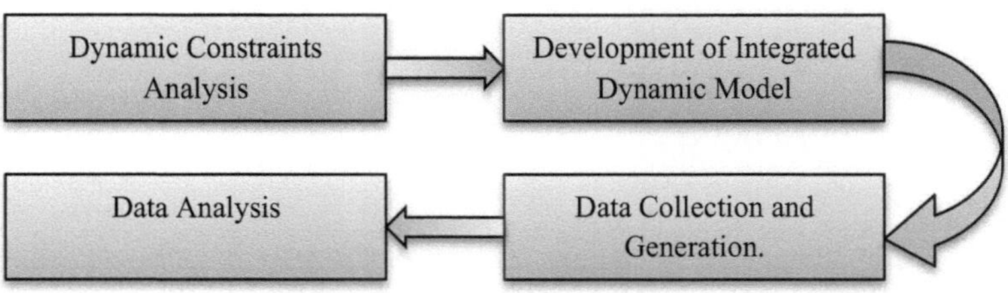

Fig. 4. Process diagram

4.1 Data Collection and Generation

In the context of computer simulations with Anaconda, data was collected to examine wheel and body movements and travel speed. This data is organized into a DataFrame that includes both input parameters and resulting simulation outputs. A systematic approach was employed, varying one parameter incrementally while holding others constant to assess the individual impact of each parameter. Fixed values were set as follows: $m_c =$ 320 kg (body mass), $m_h = 29$ kg (wheel mass), $c_p = 1680$ N·s/m (damping coefficient), $k_b = 19900$ N/m (spring stiffness), and $k_w = 165000$ N/m (tire stiffness). Parameter ranges were defined for variation, including: m_h from 10 to 100 kg, m_c from 205 to 1000 kg, c_p from 745 to 4000 N·s/m, k_b from 10995 to 60000 N/m and k_w from 104445 to 600000 N/m.

4.2 Data Analysis

In the data analysis phase, the impacts of each parameter alteration on the system's behavior are centered on. The influences of changes in mass, damping, and stiffness on wheel and body movements, as well as travel speed, are examined to distinguish patterns and noteworthy linkages. Through the use of statistical methods and visualizations, the models are emphasized, leading to a clearer understanding of the parameters that have the most significant impact on the system's displacement. This investigation is considered basic for pinpointing the most critical components that will guide future suspension system optimization efforts.

The analysis of these data is conducted using Jupyter Notebook with Anaconda, a powerful platform that integrates various data science tools and libraries. This environment leverages artificial intelligence and machine learning capabilities to enhance the accuracy and efficiency of the analysis. By utilizing advanced algorithms and computational techniques, the relationships between different parameters are more precisely identified, enabling a deeper insight into the system's behavior and facilitating more effective optimization strategies.

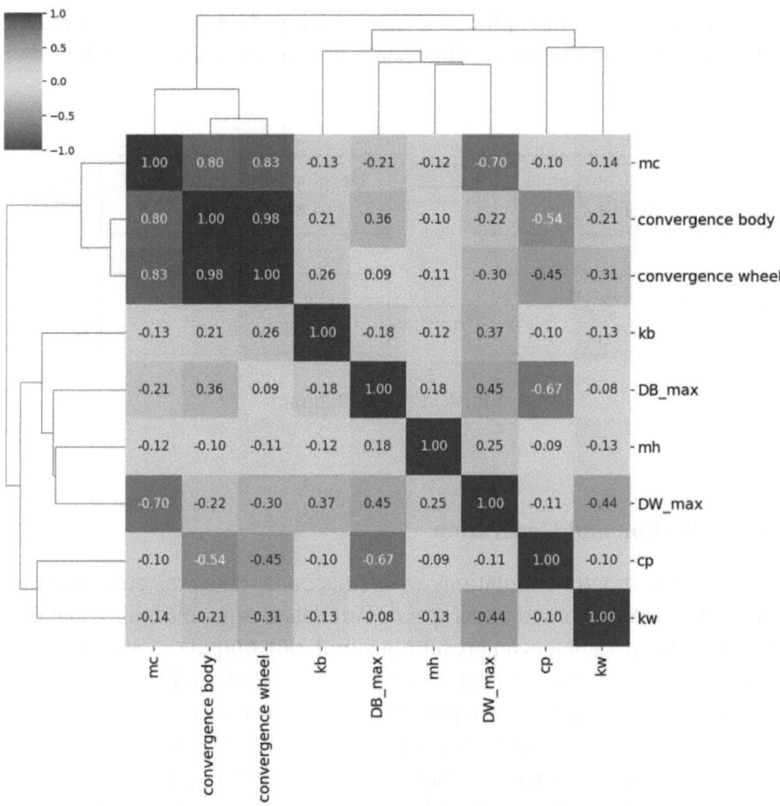

Fig. 5. Clustermap correlation between dataframes suspension system

Figure 5 presents a clustermap illustrating the correlation between various parameters of passive suspension system. The heatmap visualizes the strength of the correlation between each pair of parameters, with warmer colors indicating higher positive correlation and cooler colors indicating higher negative correlation. The dendrograms on the sides group similar parameters based on their correlation patterns.

The correlation matrix results, indicating that some variables have a notable influence on the maximum body displacement (DB_max), the maximum wheel displacement (DW_max), and the convergence of the body and wheel, while others have weaker effects. k_b has a moderate positive correlation with DW_max (0,37), suggesting that an increase in k_b leads to a moderate increase in DW_max. Conversely, c_p has a strong negative correlated with DB_max (−0.67) and with convergence body (−0,54), indicating that an increase in c_p results in a significant decrease in DB_max and convergence body. Moreover, m_c demonstrates a strong positive correlation with convergence body (0,80) and convergence wheel (0,83), meaning that an increase in m_c enhances both. However, k_b has a moderate negative correlation with convergence wheel (−0,45). Additionally, m_c is strongly negatively correlated with DW_max (−0.70), indicating that higher m_c values lead to a notable decrease in DW_max.

Figure 6 illustrates the curves of wheel displacement (BW_max) (b) and body displacement (DB_max) (a) for each parameter variation in the suspension system. The curves depict the different intervals of displacement. In the case of body displacement, the range is approximately between 0.1 m and 0.7 m, whereas for wheel displacement, it ranges between 0.18 m and 0.26 m.

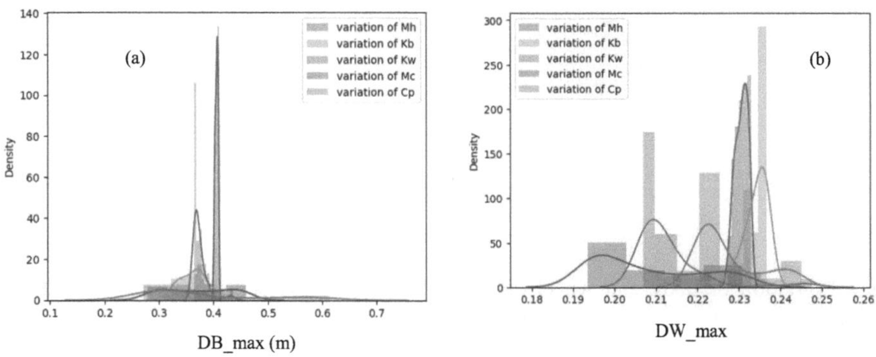

Fig. 6. Wheel (a) and body (b) displacements as a function of parameter variation

The variations of (k_w) and (m_c) significantly impact the distributions of (DW_max). While the parameters (k_w), (m_h), and (c_p) have a lesser influence.

The variations of (m_c) and (c_p) have a significant impact on the distributions of (DB_max), whereas the parameters (k_w), (k_b) and (m_h) exert a lesser influence.

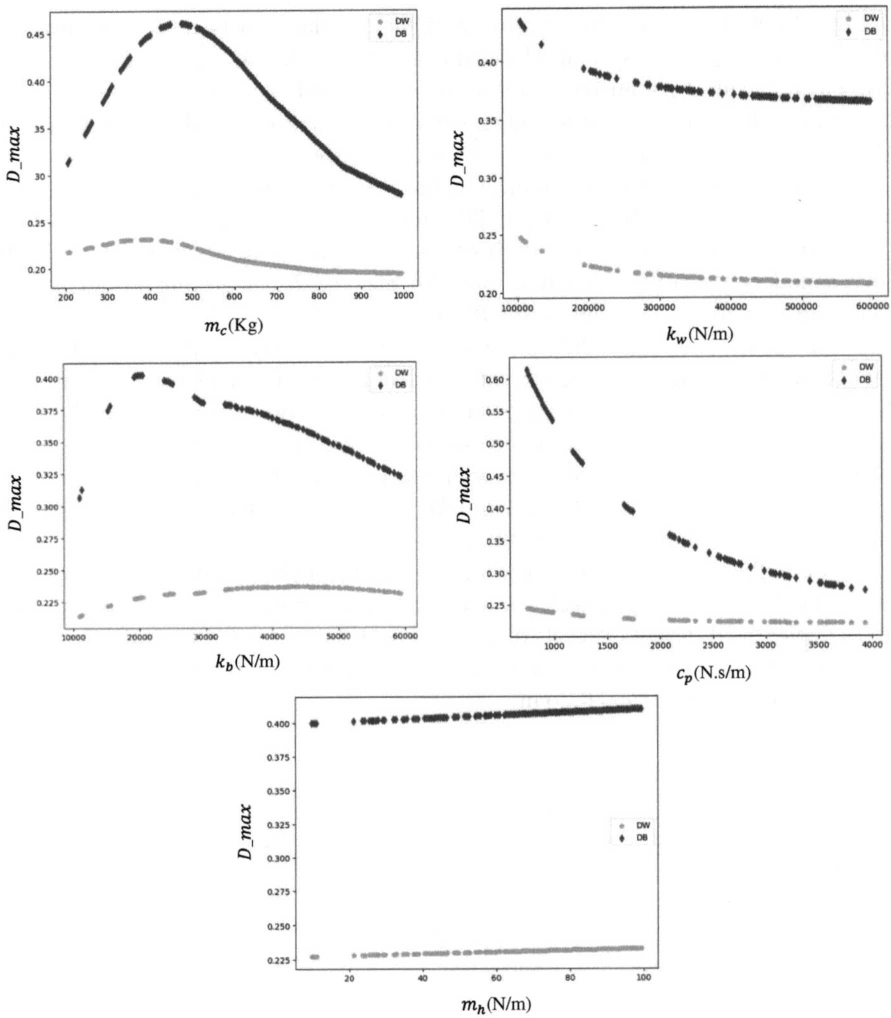

Fig. 7. Variation of body and wheel displacements for the suspension system according to $(m_h, m_c, k_w, k_b, c_p)$

The graphs in Fig. 7 illustrate the maximum displacement (D_max) of two elements, DW and DB (likely representing a wheel and a body), based on various parameters $(m_h, m_c, k_w, k_b, c_p)$ that define the mechanical properties of the system. These curves highlight how each parameter affects the system's dynamic behavior, which is essential for optimizing its design and performance.

The analyzed curves illustrate the relationships between various parameters and the maximum displacement (D_max) in a system. For the stiffness parameters k_w and k_b, an increase results in a decrease in D_max, indicating that a stiffer system limits movement. Regarding the damping coefficient c_p, D_max initially decreases before slightly increasing, highlighting the importance of optimal damping to avoid excessive rigidity. Additionally, increasing the mass m_c raises D_max due to increased inertia, making the system slower to respond to applied forces. The mass of the wheel m_h also affects D_max, although this effect is less pronounced than that of m_c. In summary, an increase in stiffness reduces movement, while an increase in mass leads to an increase in maximum displacement.

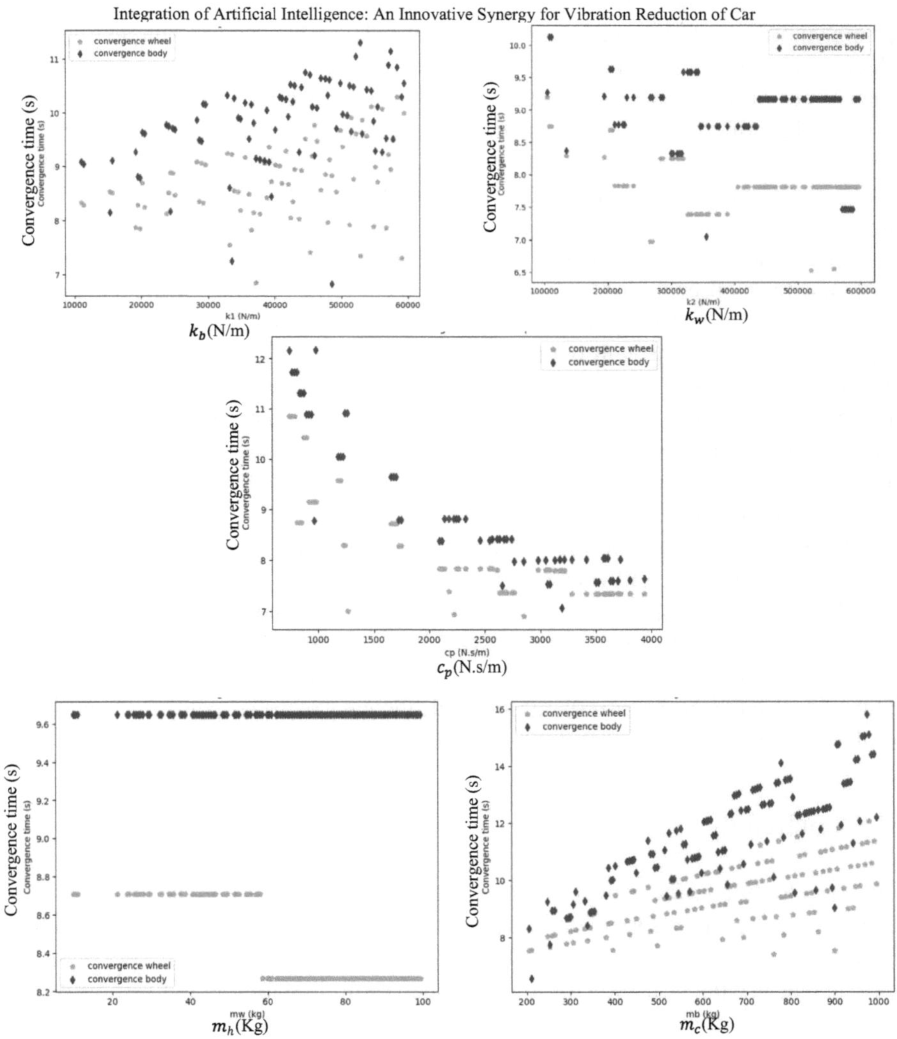

Fig. 8. Variation of wheel and body convergence times for suspension system according to $(m_h, m_c, k_w, k_b, c_p)$

The graphs in Fig. 8 illustrate the convergence times of a system consisting of the convergence time of the body and a wheel, analyzed across various parameters

$(m_h, m_c, k_w, k_b, c_p)$ The convergence times vary significantly depending on the parameters, with a notable influence from k_b and k_w, especially at high values.The parameter c_p affects the convergence time of the body more than that of the wheel, Although the parameters m_c and m_h have a less pronounced impact, a slight increase in the body's convergence time is observed with higher values of m_h. However, for lower values of m_c, the wheel's convergence time increases significantly due to system instability. Nevertheless, starting from $m_h = 57$ kg, the wheel's convergence time decreases and stabilizes for higher values of m_h, indicating an optimal balance between inertia and damping, which results in a faster and more controlled response. The behaviors of the convergence times of the body and the wheel differ based on the parameters; for instance, at certain values of c_p, the convergence time of the body increases significantly while the convergence time of the wheel remains stable. These observations may indicate that k_b and k_w represent stiffness coefficients, where high values lead to increased rigidity and shorter convergence times, while c_p could represent a damping coefficient that affects the convergence time of the body more significantly. Finally, m_c and m_h may represent masses, with an observed tendency for increased convergence times with higher masses.

In conclusion, each parameter within the suspension system has a distinct and critical impact on displacement with the damping coefficient (c_p) standing out as one of the most influential factors. An increase in (c_p) results in reduced maximum displacements and convergence times, thereby improving both stability and responsiveness. Similarly, While, vehicle mass (m_c) and spring stiffness (k_b) also impact system performance. Optimizing all the parameters is essential to achieving an effective balance between stability and ride comfort in suspension system designs.

The optimal intervals for minimizing displacements are found where k_b is between 30000 and 60000 N/m, and where m_h ranges from 57 to 100 kg for wheel convergence time. Moreover, for optimal performance in both displacement and convergence time, the most effective ranges for c_p are between 2000 and 6000 N·s/m, and for k_w between 300000 and 600000 N/m. These findings underscore the importance of precise parameter calibration to enhance the overall stability and efficiency of the suspension system. These results are crucial for achieving more precise optimization in suspension design.

5 Conclusion

This study highlights the critical parameters influencing suspension system performance, emphasizing the importance of optimizing factors such as spring stiffness, damping coefficient and vehicle mass for improved vehicle dynamics. Among these, spring stiffness is particularly vital as it directly impacts ride comfort, stability, and responsiveness. The damping coefficient significantly affects the system's ability to absorb shocks and oscillations, while vehicle mass plays a crucial role in influencing overall dynamics and maximum displacements during operation.

The integration of machine learning (ML) techniques marks a significant advancement over traditional methods, providing enhanced capabilities for suspension system optimization. The ongoing development of these AI-driven optimization results, as presented in this paper, paves the way for innovative applications that promise to further elevate suspension system performance in the future. This approach not only streamlines the design process but also opens new avenues for research and development in the automotive industry, ultimately leading to safer and more efficient vehicles.

References

Seifi, A., Hassannejad, R.: Parameters uncertainty in Pareto optimization of nonlinear inerter-based suspension system under nonstationary random road excitation. Proc. Inst. Mech. Eng. Part D J. Automob. Eng. **236**(12), 2725–2744 (2022)

Alexandru, C., Alexandru, P.: A comparative analysis between the vehicles' passive and active suspensions. Int. J. Mech. **5**(4), 371–378 (2011)

Siala, H., Mhenni, F., Choley, J.Y., Barkallah, M., Louati, J., Haddar, M.: Variational approach for robust design and sensitivity analysis of mechatronic systems. J. Chin. Inst. Eng. **43**(4), 357–364 (2020)

Mortazavi, A., Togan, V.: Sizing and layout design of truss structures under dynamic and static constraints with an integrated particle swarm optimization algorithm. Appl. Soft Comput. **51**, 239–252 (2017)

Taghavifar, H., Mardani, A., Hu, C., Qin, Y.: Adaptive robust nonlinear active suspension control using an observer-based modified sliding mode interval type-2 fuzzy neural network. IEEE Trans. Intell. Veh. **5**(1), 53–62 (2019). https://doi.org/10.1109/TIV.2019.2955365

Zhu, D., Cheng, X., Yang, L., Chen, Y., Yang, S.X.: Information fusion fault diagnosis method for deep-sea human occupied vehicle thruster based on deep belief network. IEEE Trans. Cybernet. **52**(9), 9414–9427 (2021)

Jing, Z., Chen, J., Li, X.: RBF-GA: An adaptive radial basis function metamodeling with genetic algorithm for structural reliability analysis. Reliab. Eng. Syst. Saf. **189**, 42–57 (2019)

Drehmer, L.R.C., Paucar Casas, W.J., Gomes, H.M.: Parameters optimisation of a vehicle suspension system using a particle swarm optimisation algorithm. Veh. Syst. Dyn. **53**(4), 449–474 (2015)

Xie, K., Yi, H., Hu, G., Li, L., Fan, Z.: Short-term power load forecasting based on Elman neural network with particle swarm optimization. Neurocomputing **416**, 136–142 (2020)

Theissler, A., Pérez-Velázquez, J., Kettelgerdes, M., Elger, G.: Predictive maintenance enabled by machine learning: use cases and challenges in the automotive industry. Reliab. Eng. Syst. Saf. **215**, 107864 (2021)

Barethiye, V.: Modeling and simulation of semiactive and active suspension system using quarter car model. J. Sci. Res. **16**(1), 221 (2024). https://doi.org/10.3329/jsr.v16i1.67441

Issa, M., Samn, A.: Passive vehicle suspension system optimization using Harris Hawk optimization algorithm. Math. Comput. Simul **191**, 328–345 (2022)

Modelling of a Moving Load on a Polymer Spur Gear Tooth Profile Using the FEM

Abderrahim Baccar[1,3,4](✉), Ahmed Hammami[1], Nuno Dourado[2], Ali Akrout[1], and Fakher Chaari[1]

[1] Laboratory of Mechanics, Modelling and Production, National School of Engineers of Sfax, University of Sfax, Sfax, Tunisia
abderrahim.baccar@gmail.com
[2] LABBELS –Associate Laboratory, Universidade do Minho, 4800-058 Guimarães, Portugal
[3] National School of Engineers of Tunis, University of Tunis El Manar, Tunis, Tunisia
[4] Higher Institute of Technological Studies of Mahdia, Hiboun, Tunisia

Abstract. Spur gears are essential components widely used for transmitting motion and power across various industries. In addition to traditional metal materials, polymers are increasingly utilised in gear manufacturing due to their numerous advantages. These include excellent specific mechanical and tribological properties, self-lubrication, high shock resistance, and resistance to oils and chemicals. As a result, polymer gears are commonly found in applications such as toys, mechatronic devices, computers, and laboratory instruments, where lightweight and cost-effective solutions are required. To ensure optimal performance and durability, assessing the mechanical behaviour of polymer spur gears is crucial. Several analytical and numerical techniques, including the finite element method (FEM), are widely employed to evaluate their structural response. In FEM simulations, gear teeth are often modelled as cantilever beams subjected to a moving load along the involute profile, allowing for an accurate estimation of their mechanical behaviour under different operating conditions. This study focuses on evaluating the dynamic performance of a polyamide (PA12) spur gear through a static analysis approach. Specifically, it investigates the gear mesh stiffness (GMS) by comparing different load amplitude modelling methods. The findings provide valuable insights into stiffness variations, contributing to the advancement of polymer gear design and optimization.

Keywords: Spur gear · FEM · Mesh Stiffness · LSR

1 Introduction

After examining various journals, it has been noted that the analysis of gear behaviour, typically involving the determination of the Gear Mesh Stiffness (GMS), can be carried out using several techniques, including the Finite Element Method (FEM). FEM is a numerical approach that provides approximate solutions to a range of complex issues associated with rotating machinery.

A. Hammami et al. (Eds.): MOSCOSSEE 2024, ACM 23, pp. 190–199, 2025.
https://doi.org/10.1007/978-3-032-00930-2_18

Mahr and Kissling (2014) compared tooth stiffness calculations using various analysis software. Luo et al. (2017) applied the potential energy method to define the GMS. This analytical approach was later refined by Xiong et al. (2019). Zouari et al. (2007) employed the finite element method to assess the impact of cracks in the tooth roots on the GMS. Karimpour et al. (2010) investigated the kinematics of polymer cylindrical gears using both the FEM and an analytical approach. Chakroun et al. (2022, 2023) introduced a novel approach for calculating the GMS, which is based on the generalized Maxwell model and incorporates the viscoelastic properties of the polymer. On the other hand, Tsai et al. (1997), Kodeeswaran et al. (2017), and Cathelin et al. (2013) focused on analysing transmission error without considering GMS variations.

In this research, the dynamic behaviour of a basic polymer gear system is examined using the Finite Element Method (FEM), with an emphasis on static analysis. The proposed method treats the gear tooth as a cantilever beam with a moving contact load. Various approaches exist in the literature for modelling load amplitude. This chapter aims to compare the GMS results obtained using different methods of load amplitude modelling.

2 Creation of the Model

To model the behaviour of a plastic gear using Finite Element Method (FEM), the initial step is to develop a digital representation that incorporates the key parameters specific to the case study. These parameters include the physical model's geometry, applied loads, mesh elements, and material properties. A model featuring only three teeth, as depicted in Fig. 1, was first constructed using SOLIDWORKS software and subsequently imported into simulation software. The model, designed to be compatible with both the wheel and the pinion, was created based on the specifications outlined in Table 1.

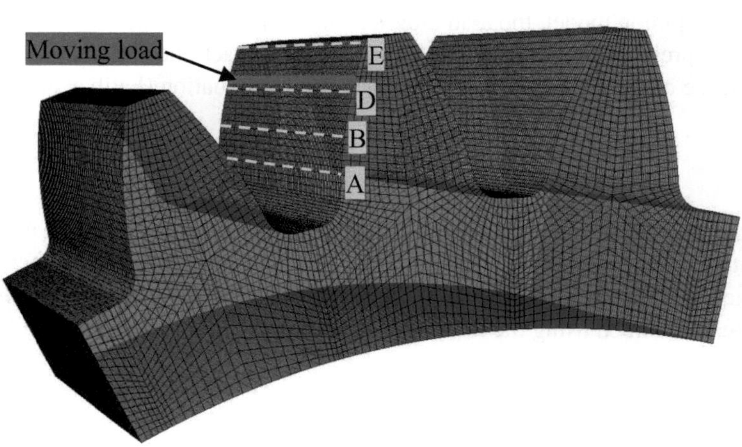

Fig. 1. Gear model.

Table 1. Gears features.

Parameters	Pinion (1)	Wheel (2)
Module	$m = 2$	
Teeth number	$Z_1 = 23$	$Z_2 = 30$
Pitch circle	$d_1 = 46\ mm$	$d_2 = 60\ mm$
Addendum circle	$d_{a1} = 50\ mm$	$d_{a2} = 64\ mm$
Base circle	$d_{b1} = 43.23\ mm$	$d_{b2} = 56.38\ mm$
Pressure angle	$\alpha_n = 20°$	
Face width	$b = 17\ mm$	
Contact ratio	$c = 1.6\ mm$	
Rim thickness	$r_1 = 3\ mm$	$r_2 = 3.5\ mm$

The next step in the analysis involves specifying the boundary conditions and applied loads. In this study, a load F is applied solely to the central tooth profile, as a normal pressure on the contact surface, which can move along the contact zone (AE) (Fig. 1). Meanwhile, the rim of the model is fixed in place. Points 'A' and 'E' denote the intersections of the addendum circles of the wheel and pinion with the line of action (Fig. 1). The contact segment between the pinion (1) and wheel (2) is defined using the relationship described by Fatih et al. (2017).

$$|AE| = \frac{1}{2}\left|\sqrt{d_{a1}^2 - d_{b1}^2} + \sqrt{d_{a2}^2 - d_{b2}^2} - (d_1 + d_2)\sin(\alpha_n)\right| \tag{1}$$

This implies that, in the wheel model, the load moves from point 'E' to point 'A', while in the pinion model, the load moves from point 'A' to point 'E'. The contact ratio c, which represents the number of tooth pairs in contact and influences the dynamic performance of the system, is given by the following equation (Fatih et al., 2017),

$$c = \frac{|AE|}{\pi \cdot m \cdot \cos(\alpha_n)} \tag{2}$$

In our case, with a transmission contact ratio ranging from 1 to 2 (Table 1), when the contact occurs at segments (AB) or (DE), two pairs of teeth are in contact. However, at segment (BD), only a single pair of teeth is in contact. The positions of points B and D can be determined using the following relationship (Fatih et al., 2017).

$$|AB| = |DE| = |AE| - \pi \cdot m \cdot \cos(\alpha_n) \tag{3}$$

When one tooth makes contact with another, two additional teeth are already in contact at segment (AB). At point B, the contact between the two teeth ends, so during segment (BD), only one pair remains in contact. In the final phase of contact, at segment (DE), a new pair of teeth engages. This meshing process continues cyclically. During each meshing cycle, the load F transfers from one elementary surface to another over time.

According to Lee (2009), a structured mesh using hexahedral elements is most suitable for modelling solid elasticity in three-dimensional problems. For this study, a structured first-order linear hexahedral mesh with 8 nodes was generated. At an environmental temperature of around 25 °C, the PA12 thermoplastic material has an elastic modulus of 1650 MPa and a Poisson's ratio of 0.3, as noted by Baccar et al. (2023). This polymer is chosen for this study.

3 Mesh Stiffness Calculation

Since rigidity reflects a body's ability to resist deformation when subjected to an applied load, calculating the GMS requires determining the deflection of the contacting teeth at each point in time. The stiffness of an individual tooth is defined as the ratio of the total transmitted load F to the deflection Δ of the tooth at the load application point 'I', measured along the direction of the load's action (Chaari et al., 2009).

$$K_{Tooth} = F/\Delta \tag{4}$$

The stiffness of a single tooth pair in contact, K_m, is determined by treating the stiffness of the two contacting teeth K_1 (the pinion tooth stiffness) and K_2 (the wheel tooth stiffness) as the equivalent stiffness of two springs connected in series.

$$K_m = (K_1 \times K_2)\Big/(K_1 + K_2) \tag{5}$$

During the meshing phase, either one or two tooth pairs are in contact. Therefore, the total stiffness, GMS, is calculated by summing the individual stiffnesses of the contacting pairs. As a result, the GMS fluctuates between one and two pairs of teeth in contact (Chaari et al., 2009).

$$K_e = \sum K_m(i) \tag{6}$$

4 Load Modelling Methods

During operation, the theoretical load F applied to the tooth profile is directly proportional to the input torque C_1 and the pitch diameter of the pinion d_1, as defined by:

$$F = \frac{2C_1}{d_1.\cos(\alpha_n)} \tag{7}$$

When the contact is localized to the segment (BD), where only a single pair of teeth is in contact, the entire load is applied to the profiles of the two contacting teeth. However, when two pairs are in contact, as seen in segments (AB) or (DE), the load is distributed between the two pairs of teeth. The literature presents two methods for determining how the load is shared between the different contacting pairs.

The first method, CUR (Uniformly Distributed Load), assumes that the load is constant and evenly distributed across the contacting pairs. Used by Chaari et al. (2009) and

Fatih et al. (2017), this method requires applying the total load to determine the GMS at any point along the contact segment (AE), with the assumption that the entire load is carried by a single pair of teeth in contact.

The second method models the contact load using a variable load-sharing coefficient known as the Load Sharing Ratio (LSR). According to Sánchez et al. (2017), the LSR for a straight cylindrical gear (with $1 < c \leq 2$) is defined as a purely geometric parameter, which can be expressed using the following relations:

$$LSR(\xi_I) = 0.36 - \tfrac{0.28}{c-1}(\xi_I - \xi_{inn}) \text{ For } \xi_{inn} \leq \xi_I \leq \xi_{inn} + c - 1 \tag{8}$$

$$LSR(\xi_I) = 1 \text{ For } \xi_{inn} + c - 1 \leq \xi_I \leq \xi_{inn} + 1 \tag{9}$$

$$LSR(\xi_I) = 0.36 - \tfrac{0.28}{c-1}(\xi_I - \xi_{inn} - c) \text{ For } \xi_{inn} + 1 \leq \xi_I \leq \xi_{inn} + c \tag{10}$$

The profile parameter ξ_I, which describes the contact point 'I' along the contact segment (AE), is defined for the pinion using the following relation:

$$\xi_I = \frac{Z_1}{2\pi}\sqrt{\frac{d_I^2}{d_b^2} - 1} \tag{11}$$

The contact point 'I' of the tooth's involute profile, defined by its circle diameter d_I, moves between points "A" and "E" (see Fig. 1). To determine the load model throughout a complete cycle, the load amplitude is multiplied by the LSR coefficient, as shown in Eq. (8), to represent the load during the cycle phase when double pairs are in contact (zone AB, Fig. 2). Equation (9) applies to the load in zone BD, where a single pair is in contact (Fig. 2), and Eq. (10) is used for the load in the final phase of the cycle. The variations of the LSR and CUR coefficients over a single cycle are depicted in Fig. 2.

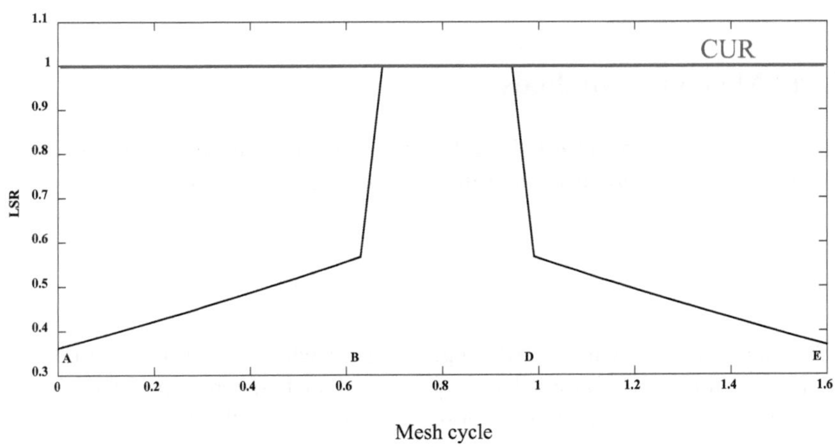

Fig. 2. Load coefficient.

5 Results

A static procedure was used in the FEM analysis to examine various stiffness types under a 100 N load intensity. Both the CUR and LSR methods were employed to determine and compare the stiffness results, allowing for an assessment of each method's performance. As illustrated in Fig. 3, the wheel's stiffness calculated using the CUR method ranges from 1769 N/mm to 7876 N/mm, while the stiffness calculated using the LSR method spans from 1701 N/mm to 7318 N/mm by the end of the cycle.

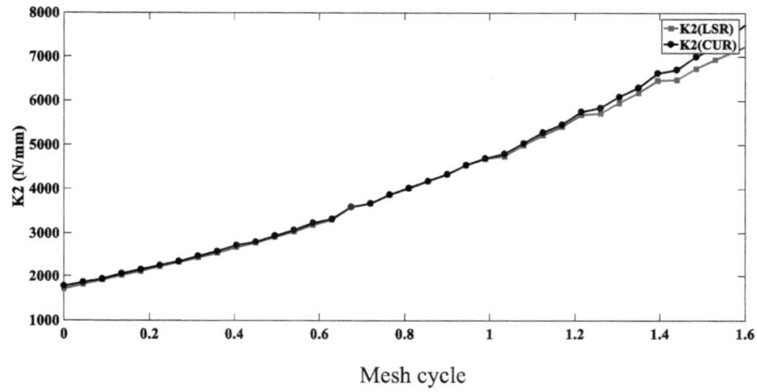

Fig. 3. Wheel tooth stiffness.

For the pinion stiffness shown in Fig. 4, applying the CUR method produces a decreasing curve that ranges from 7635 N/mm to 1745 N/mm. In contrast, when the LSR method is considered, the red curve fluctuates between 7076 N/mm and 1680 N/mm throughout the meshing cycle.

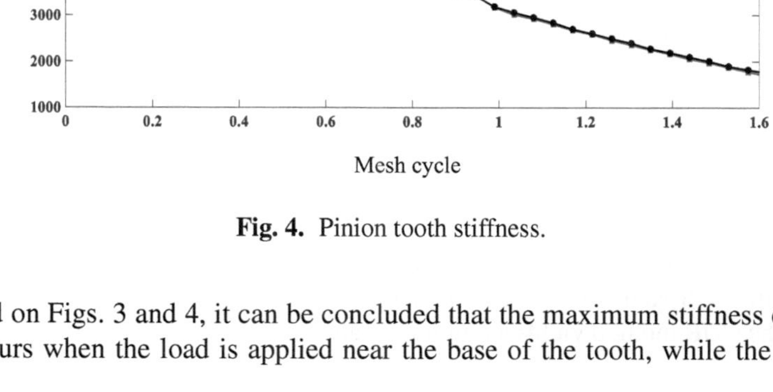

Fig. 4. Pinion tooth stiffness.

Based on Figs. 3 and 4, it can be concluded that the maximum stiffness of a single tooth occurs when the load is applied near the base of the tooth, while the minimum stiffness is observed when the contact is localized near the tooth's head. Figures 5 and 6 illustrate the stiffness of a single pair of teeth in contact K_m and the GMS using the LSR and CUR methods. With the LSR method, the stiffness curve of a single pair (Km) follows a parabolic shape, reaching a maximum value of 1965 N/mm at the midpoint of the cycle and a minimum value of 1366 N/mm. In contrast, the CUR method yields a stiffness curve with a maximum of 1964 N/mm at the cycle's midpoint and a minimum value of 1428 N/mm.

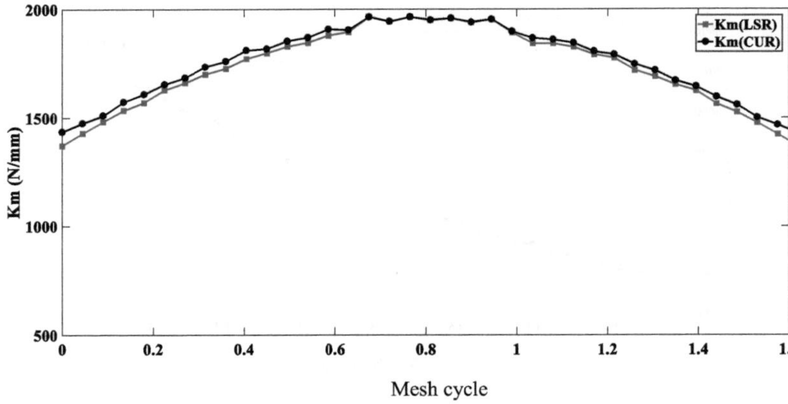

Fig. 5. Single pair mesh stiffness K_m.

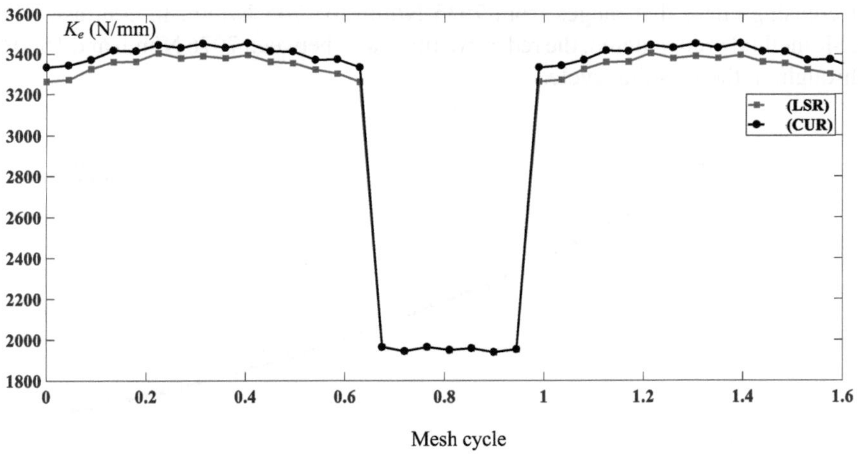

Fig. 6. GMS.

At the level shown in Fig. 6, the LSR method (represented by the red curve) achieves a maximum GMS value of 3405 N/mm and a minimum value of about 1938 N/mm. In contrast, the CUR method yields a minimum value of 1939 N/mm and a maximum value of 3405 N/mm. Looking at the previous four figures (Figs. 3, 4, 5, and 6), three distinct zones—(AB), (BD), and (DE)—can be identified.

In the zones where two pairs of teeth are in contact, namely (AB) and (DE), the LSR model shows a decrease in stiffness compared to the curve produced by the CUR method. However, in the (BD) zone, where only one pair of teeth is in contact, both curves (LSR and CUR) align perfectly. Notably, in the (AB) and (DE) zones, the reduction in stiffness is more significant for both the K_m and GMS curves. Additionally, the difference between the two curves becomes apparent primarily at the start of the cycle (Zone AB) for K_1 and at the end of the cycle (Zone DE) for K_2. It is important to calculate the percentage error between the curves derived from the LSR and CUR methods for the various scenarios. This error can be determined using the following formula:

$$er(K) = \frac{K(CUR) - K(LSR)}{K(CUR)} \tag{12}$$

The error between the two methods for the stiffness of a single pair, K_m, can reach a maximum of approximately 4.49%, while the error for the GMS ranges from 0% to 2.15% (as shown in Fig. 7).

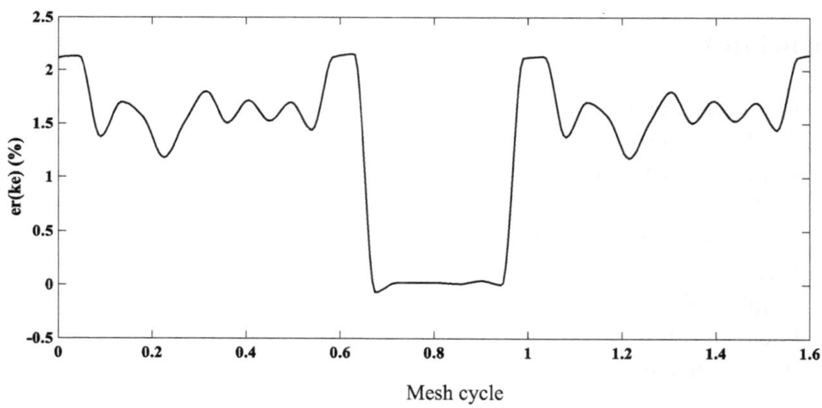

Fig. 7. Error curve.

6 Conclusion

This study primarily focused on analysing the engagement stiffness (K_e) of cylindrical gear trains made from polymers under various operating conditions, taking into account the mechanical properties of the selected material. Finite Element Method (FEM) simulations were used to evaluate how different load amplitude modelling approaches (CUR and LSR) impact stiffness variations. The key findings are as follows:

- The engagement stiffness varies based on the number of teeth pairs in contact and can be derived from the stiffness of a single pair.

- The CUR method results in higher stiffness compared to the LSR method in regions where two pairs of teeth are in contact.
- When only a single pair of teeth is in contact, the stiffness values (K_e) obtained from both the LSR and CUR methods are virtually identical.
- Since the load is typically distributed among multiple pairs in contact in practical applications, the LSR method is more effective for simulating the gear's overall behaviour.

Acknowledgements. This work is supported by Tunisian-Portuguese project collaboration between Mechanics, Modelling and Production laboratory (LA2MP) and LABBELS –Associate Laboratory.

Tribute. The authors would like to pay tribute to Professor Fakher CHAARI and appreciate his guidance and contribution to this work. May his soul rest in peace.

Appendix

Nomenclature

FEM	Finite element method
LSR	Load sharing ratio
CUR	Uniformly distributed load
er	Stiffness error
K_e	Gear mesh stiffness (GMS)
m	Module
Z	Number of teeth
d	Pitch circle
d_a	Addendum circle
d_b	Base circle
b	Face width
c	Contact ratio
α_n	Pressure angle
r	Rim thickness
F	Load
Δ	Tooth deflection
K_{Tooth}	Tooth stiffness
K_m	Single pair stiffness
C	Torque
ξ_I	Profile parameter

References

Baccar, et al.: Determination of the crack propagation effect on the mesh stiffness for a polymer spur gear tooth using the extended finite element method. J. Braz. Soc. Mech. Sci. Eng. **45**, 286 (2023)

Chaari, et al.: Analytical modelling of spur gear tooth crack and influence on gearmesh stiffness. Eur. J. Mech. A. Solids **28**, 461 (2009)

Cathelin, et al.: Experimental and numerical study of a loaded cylindrical PA66 gear. J. Mech. Des. **135**, 4 (2013)

Chakroun, et al.: Gear mesh stiffness of polymer-metal spur gear system using generalized Maxwell model. Mech. Mach. Theory **175**, 104934 (2022)

Chakroun, et al.: Numerical and experimental study of the dynamic behaviour of a polymer-metal worm drive. Mech. Syst. Signal Process. **193**, 110263 (2023)

Fatih, et al.: An improved numerical method for the mesh stiffness calculation of spur gears with asymmetric teeth on dynamic load analysis. Adv. Mech. Eng. (2017). https://doi.org/10.1177/1687814017721856

Karimpour, et al.: A kinematic analysis of meshing polymer gear teeth. Proc. Inst. Mech. Eng. Part L **224**(3), 101–115 (2010)

Kodeeswaran, et al.: Effect of strain rate on bending and transmission characteristics of injection molded polyamide 66 spur gears. Proc. Inst. Mech. Eng. Part L J. Mater. Des. Appl. **233**(6), 1145–1155 (2017)

Lee: Non-Linear Contact Analysis of Meshing Gears. Thesis Presented to The Faculty of California Polytechnic State University, San Luis Obispo (2009)

Luo, et al.: Effects of gear center distance variation on time varying mesh stiffness of a spur gear pair. Eng. Fail. Anal. **75**, 37–53 (2017)

Mahr, Kissling: Comparison between Different Commercial Gear Tooth Contact Analysis Software Packages. KISSsoft AG (2018)

Sánchez, et al.: Approximate equations for the meshing stiffness and the load sharing ratio of spur gears including hertzian effects. Mech. Mach. Theory **109**, 231–249 (2017)

Tsai, et al.: A method for calculating static transmission errors of plastic spur gears using FEM evaluation. Finite Elem. Anal. Des. **27**(4), 345–357 (1997)

Xiong, et al.: Research on the influence of backlash on mesh stiffness and the nonlinear dynamics of spur gears. Appl. Sci. **9**(5), 1029 (2019)

Zouari, et al.: Three-dimensional analyses by finite element method of a spur gear: effect of cracks in the teeth foot on the mesh stiffness. J. Fail. Anal. Prev. **7**(6), 475–481 (2007)

A Generalised Stiffness Model for Incorporating Multiple Spur Gear Tooth Faults - Part 1: Stiffness Modelling

Luke van Eyk(✉), P. Stephan Heyns, and Stephan Schmidt

Centre for Asset Integrity Management, University of Pretoria, Pretoria, South Africa
luke.vaneyk@up.ac.za

Abstract. Gear mesh stiffness (GMS) is a critical parameter in modelling gearbox vibration response using lumped mass methods. Existing GMS models in the literature are often fragmented, limiting their application to broader condition-based maintenance strategies. This chapter introduces a generalised spur GMS model based on the potential energy method, which can integrate multiple fault types within a unified framework. The framework facilitates the computation of time-varying stiffness, incorporating bending, shear, axial, Hertzian, and fillet foundation effects, ensuring a physically consistent stiffness estimation. This is the first gear mesh stiffness model that simultaneously models multiple damage types within a single unified framework, which is advantageous for the gear condition-based maintenance community. This work represents the first of two parts. This first part is concerned with generalised stiffness modelling, with the second part addressing fault modelling. In addition to a rigorous mathematical description, the proposed methodology is implemented in an open-source Python package (https://github.com/LukevanEyk/Gearbox-Simulator), which allows users to input spur gear parameters, introduce faults, and analyse the resulting stiffness variations. By providing an accessible and computationally efficient approach, this framework supports academic research and industrial applications in gearbox diagnostics, condition monitoring, and predictive maintenance.

Keywords: Generalised stiffness model · Potential energy method · Spur gear fault modelling · Time-varying mesh stiffness

1 Introduction

Gearboxes are integral to many industrial applications, including wind turbines, mills, trucks, and critical systems like helicopters (Chaari et al. 2008, Liu et al.2021). These gearboxes operate in harsh environments and are subject to failure, primarily due to gear tooth faults (Liang et al. 2016, Liu et al. 2021). Tooth faults often reduce the stiffness of gear teeth, leading to increased deformation of the damaged tooth during meshing. The increased deformation manifests as a reduced gear meshing stiffness (GMS) between a tooth pair, making GMS

A. Hammami et al. (Eds.): MOSCOSSEE 2024, ACM 23, pp. 200–215, 2025.
https://doi.org/10.1007/978-3-032-00930-2_19

modelling an essential focus within gearbox research (Meng et al. 2020). The potential energy method (PEM), developed by Yang & Lin (1987), is widely used for GMS modelling due to its accuracy, often comparable to finite element methods (Luo et al. 2018).

The PEM has seen continuous evolution, with significant contributions addressing axial, bending, shear and fillet foundation stiffnesses (Tian 2004, Sainsot et al. 2004, Ma et al. 2016). Parallel to these advancements, various fault models have emerged as shown in the reviews of Mohammed & Rantatalo (2020) and Van Eyk (2021). Despite significant advancements in GMS and fault modelling, several challenges remain. The literature on GMS and fault modelling is fragmented, exhibiting inconsistencies in stiffness terminology, notation and modelling techniques - particularly within the potential energy method (PEM) (Van Eyk et al. 2022). Furthermore, existing studies typically address individual fault types, with limited attempts to integrate different fault models into a unified framework (Yang et al. 2021).

This work proposes a unified GMS and fault modelling framework that offers several key advantages. First, it provides a standardised basis for comparing different models and methods. Second, it enables the simulation of multiple fault types on different gear teeth, an approach not previously explored. Finally, such a framework supports the creation of synthetic datasets with diverse fault types, which are essential for training hybrid condition monitoring systems (Liu et al. 2021, Sobie et al. 2018). This chapter represents the first of two parts and focuses on the GMS modelling methodology, with the second part focussing on the fault modelling methodology. This chapter therefore contributes a GMS modelling framework that supports multiple fault types across gear teeth. Furthermore, the framework has been published as open source Python code which is publically available at: https://github.com/LukevanEyk/Gearbox-Simulator. This code allows users to input spur gear geometry and fault data and pair this with a gearbox model to obtain the resulting GMS, fault visualisations and solved gearbox dynamics. The latter feature facilitates the generation of simulated gearbox data for machine learning. The remainder of the chapter describes the GMS modelling approach (Sect. 2), and concludes with recommendations in Sect. 3.

2 Generalised Gear Mesh Stiffness Modelling

Physics-based models play a vital role in gearbox condition monitoring, enabling both diagnostic and prognostic applications. Accurate models that capture the physics of meshing gears are particularly valuable, as meshing behaviour is the primary source of excitation within a gearbox. Fault modelling further enhances the utility of these models, as damaged gears directly alter gear meshing stiffness (GMS), an essential metric studied extensively in the literature. The former is considered here, with the latter considered in a subsequent chapter.

Two factors significantly influence GMS: the number of simultaneous tooth pairs in contact and the variation in force angle and location along the line of action. During gear rotation, load distribution shifts from double tooth pairs to a single pair, causing primary stiffness fluctuations. Additionally, the contact force angle and position relative to the gear tooth vary throughout the rotation, causing smaller, secondary stiffness variations. Modelling these interactions is crucial for understanding meshing behaviour. This work employs the potential energy method (PEM), introduced by Yang & Lin (1987), to model these effects analytically. PEM provides a balance between computational efficiency and accuracy compared to finite element methods (FEM) (Van Eyk et al. 2022). To address the limitations of existing models, a generalised stiffness model (GenSM) is proposed that integrates GMS and fault modelling into a unified framework. The GenSM operates at the tooth level, allowing for the simulation of multiple fault types with varying intensities on different teeth. Unlike prior approaches, this framework enables the simultaneous study of various faults on the same gear (Van Eyk et al. 2022). To aid the explanation of the proposed GenSM, a detailed summary of the modelling procedure is given in Fig. 1.

2.1 Gear-Specific Parameters and Gear Mesh Period Discretisation

The GenSM utilises key gearbox parameters related to gear geometry, interaction angles, and machining assumptions. Table 1 summarises the quantities required for modelling within the proposed framework, with default values commonly used in the literature indicated by asterisks (*). Subsequent gearbox parameters can be derived from these fundamental parameters.

The first derived modelling parameter is the pinion meshing period (θ_t). The meshing period refers to the angular distance between the start and end of meshing for a given pinion tooth, N. This parameter is defined as:

$$\theta_t = 2\theta_d + \theta_s \tag{1}$$

where the double tooth meshing angle (θ_d) and the single tooth meshing angle (θ_s) is defined by Tian (2004) as:

$$\begin{aligned}\theta_d = &\tan\left(\arccos\frac{Z_1 \cos\alpha_0}{Z_1+2}\right) - \frac{2\pi}{Z_1} \\ &- \tan\left[\arccos\frac{Z_1\cos\alpha_0}{\sqrt{(Z_2+2)^2 + (Z_1+Z_2)^2 - 2(Z_2+2)(Z_1+Z_2)\cos\left(\arccos\left(\frac{Z_2\cos\alpha_0}{Z_2+2}\right) - \alpha_0\right)}}\right]\end{aligned} \tag{2}$$

$$\theta_s = \frac{2\pi}{Z_1} - \theta_d \tag{3}$$

The angular range of contact for each tooth can therefore represented by:

$$\theta_i \in [0, \theta_t] \tag{4}$$

where a unique meshing stiffness value is calculated for a given pinion tooth N and a given angular value θ_i in the set $[0, \theta_t]$.

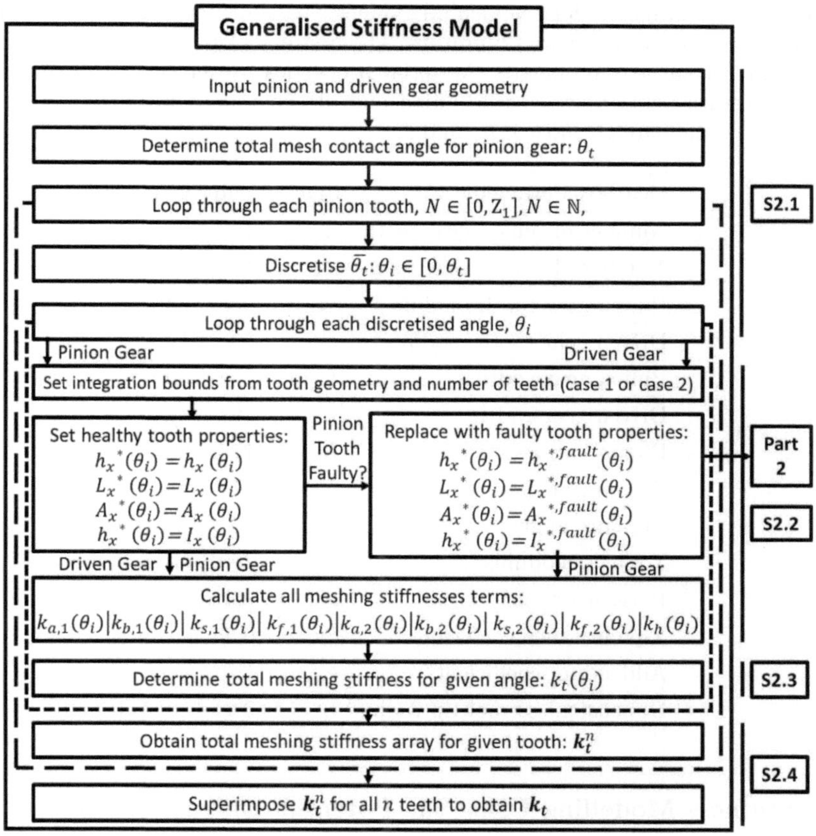

Fig. 1. Blocks on the right-hand side indicate which sections pertain to which modelling steps. Pinion gear angles are divided into Z_1 groups of teeth, each in mesh for θ_t rad. Each meshing cycle is subdivided into individual meshing angles, $\theta_i \in [0, \theta_t]$ per tooth. For a given tooth and angular position, the relevant case and corresponding integrals are selected. If required, faults are introduced (see part 2). After fault introduction, stiffness terms for the pinion (subscript 1) and driven gear (subscript 2) are calculated. The stiffness terms are combined to obtain the total meshing stiffness (k_t) for a single angle of a single tooth. This process is repeated over all angles to obtain a set of stiffnesses, $\mathbf{k_t^{(n)}}$. Finally, all Z_1 teeth are superimposed to obtain $\mathbf{k_t}$, the total meshing stiffness for any given rotation angle of the pinion gear.

Table 1. Key parameters for the GenSM. Values with an asterisk indicate commonly assumed or re-used values across literature. The first eight quantities refer to physical pinion and driven gear properties. Material properties are given by E and ν. The relative pressure angle between the pinion and driven gear is given by α_0. Finally, tooling parameters are given by c, h_a and $\bar{r}_c$.

Gearbox Parameter	Symbol	Value	Unit
Number of pinion gear teeth	Z_1	19	-
Number of driven gear teeth	Z_2	19	-
Pinion gear tooth module	m_1	3.2	mm
Driven gear tooth module	m_2	3.2	mm
Pinion gear width	L_1	38	mm
Driven gear width	L_2	38	mm
Pinion hub radius	R_{h1}	17.5	mm
Driven gear hub radius	R_{h2}	17.5	mm
Pressure angle*	α_0	20	°
Young's modulus*	E	200	GPa
Poisson's ratio*	ν	0.3	-
Tip clearance coefficient*	c	0.25	-
Addendum coefficient*	h_a	1	-
Dimensionless tool tip radius*	$\bar{r}_c$	0.2	-

2.2 Stiffness Modelling Using the Potential Energy Method

For each pinion tooth (N) and each discretised meshing angle (θ_i), the GMS is determined using the PEM. Key parameters, including the base radius (R_b) and root radius (R_r) for the pinion and driven gear, are derived from standard spur gear geometry (Ma et al. 2014):

$$R_{b_k} = 0.5 m_k Z_k \cos \alpha_0 \tag{5}$$

$$R_{r_k} = 0.5 m_k Z_k - (h_a + c) m_k \tag{6}$$

where $k = 1$ for the pinion gear and $k = 2$ for the driven gear. Default tooling parameters $h_a = 1$ and $c = 0.25$ are used, following Ma et al. (2014) and (Wan et al. 2014).

The PEM models a gear tooth as a non-uniform cantilever beam, as illustrated in Fig. 2. Key parameters include the contact force (F) at angle α_1, the length (x) and height (h_x) at an arbitrary point, and the corresponding tooth width (L_x), cross-sectional area (A_x), and moment of inertia (I_x). At the force contact point, these parameters are denoted as d and h.

The PEM calculations are categorised into three components:

1. Beam stiffness terms: Derived using cantilever beam theory to account for bending, shear, and axial stiffness.

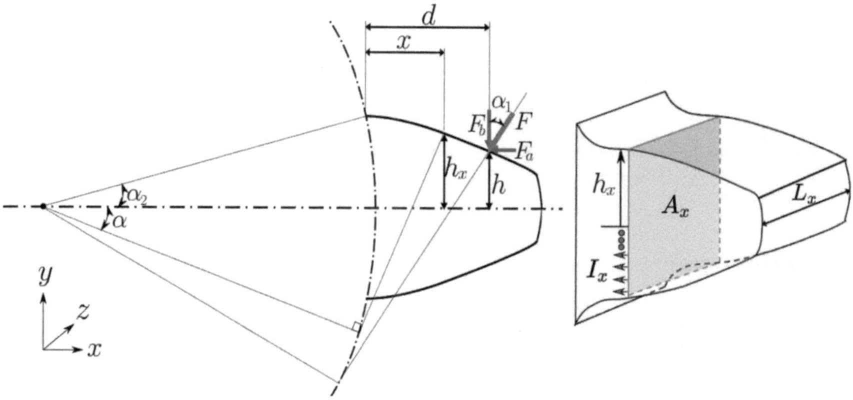

Fig. 2. Key parameters for modelling a gear tooth.

2. Hertzian stiffness term: Accounts for contact deformation between the gear teeth.
3. Fillet foundation stiffness term: Models deflections at the tooth base.

These terms collectively define the total meshing stiffness, as detailed in subsequent sub-sections.

Beam Stiffness Terms (k_a, k_b, k_s): Using the PEM, axial (k_a), bending (k_b), and shear (k_s) stiffness terms are derived by modelling the gear tooth as a cantilever beam (Tian 2004):

$$U_a = \frac{F^2}{2k_a} = \int_0^d \frac{F_a^2}{2EA_x} dx \tag{7}$$

$$U_b = \frac{F^2}{2k_b} = \int_0^d \frac{[F_b(d-x) - F_a h]^2}{2EI_x} dx \tag{8}$$

$$U_s = \frac{F^2}{2k_s} = \int_0^d \frac{1.2F_b^2}{2GA_x} dx = \int_0^d \frac{1.2(1+\nu)F_b^2}{2EA_x} dx \tag{9}$$

where E, ν, and G are the modulus of elasticity, Poisson's ratio, and shear modulus, respectively. Forces F_a and F_b depend on the contact angle α_1:

$$F_b = F\cos\alpha_1 \tag{10}$$

$$F_a = F\sin\alpha_1 \tag{11}$$

where α_1 will be defined shortly. Integration in the $x-y$ domain is challenging, so an angular coordinate system based on α is used (Yang & Lin 1987). The contact angle α_1 varies with the meshing cycle angle θ_i and differs for the pinion and driven gear (Tian 2004):

For the pinion:

$$\alpha_1(\theta_i) = \theta_i - \frac{\pi}{2Z_1} - (\tan\alpha_0 - \alpha_0) + \tan\left[\arccos\left(\frac{Z_1\cos\alpha_0}{\sqrt{(Z_2+2)^2+(Z_1+Z_2)^2-2(Z_2+2)(Z_1+Z_2)\cos\left(\arccos\left(\frac{Z_2\cos\alpha_0}{Z_2+2}\right)-\alpha_0\right)}}\right)\right] \tag{12}$$

For the driven gear:

$$\alpha_1(\theta_i) = \tan\left(\arccos\left(\frac{Z_2\cos\alpha_0}{Z_2+2}\right)\right) - \frac{\pi}{2Z_2} - (\tan\alpha_0 - \alpha_0) - \frac{Z_1}{Z_2}\theta_i \tag{13}$$

Although it is shown that α_1 is only a function of the pinion rotation angle, the dependency on θ_i is not explicitly shown in subsequent discussions, i.e. $\alpha_1(\theta_i)$ is written as α_1. The half-tooth angle α_2 is given by (Meng et al. 2020):

$$\alpha_2 = \frac{\pi}{2Z} + \tan\alpha_0 - \alpha_0 \tag{14}$$

One of two cases is applicable, depending on the ratio of calculated root and base radii. Case one is relevant where the root radius is smaller than the base radius. Otherwise, case two is relevant. The geometry of each of these cases may be seen in Fig. 3.

Case 1: $R_r \leq R_b$: For this case, α_3 may be found from[1]

$$\alpha_3 = \arcsin\left(\frac{R_b\sin\alpha_2}{R_r}\right) \tag{15}$$

which aims to locate the angle at which the tooth profile reaches the root circle (Meng et al. 2020). The following transformation equations can be defined to convert between the angular and $x - y$ coordinate systems (Meng et al. 2020):

$$h = R_b[(\alpha_1 + \alpha_2)\cos\alpha_1 - \sin\alpha_1] \tag{16}$$

$$d = R_b[(\alpha_1 + \alpha_2)\sin\alpha_1 + \cos\alpha_1] - R_r\cos\alpha_3 \tag{17}$$

$$d_1 = R_b\cos\alpha_2 - R_r\cos\alpha_3 \tag{18}$$

$$x = \begin{cases} x, & \text{if } 0 \leq x \leq d_1 \\ R_b[(\alpha - \alpha_2)\sin\alpha + \cos\alpha] - R_r\cos\alpha_3, & \text{if } d_1 < x \leq d \end{cases} \tag{19}$$

$$h_x = \begin{cases} h_{x_1} = R_b\sin\alpha_2, & \text{if } 0 \leq x \leq d_1 \\ h_{x_2} = R_b[(\alpha_2 - \alpha)\cos\alpha + \sin\alpha], & \text{if } d_1 < x \leq d \end{cases} \tag{20}$$

$$A_x = 2h_xL_x \tag{21}$$

$$I_x = \frac{1}{12}(2h_x)^3L_x = \frac{2}{3}h_x^3L_x \tag{22}$$

[1] This equation is only valid if for a straight fillet estimation.

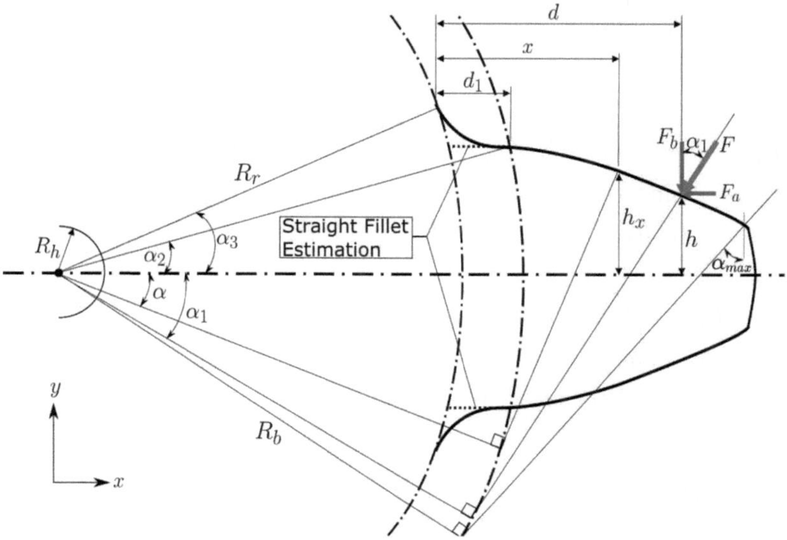

(a) Case 1: $R_r \leq R_b$. The x-direction length between the point where the root and base radius intersect the tooth is labelled d_1. The hub radius is given by R_h. The angle between the intersection point of the root radius and the tooth gear and the centreline is given by α_3. Finally, α_{max} indicates the point along the tooth where the maximum α value will be obtained during integration. Figure adapted from Van Eyk et al. (2022).

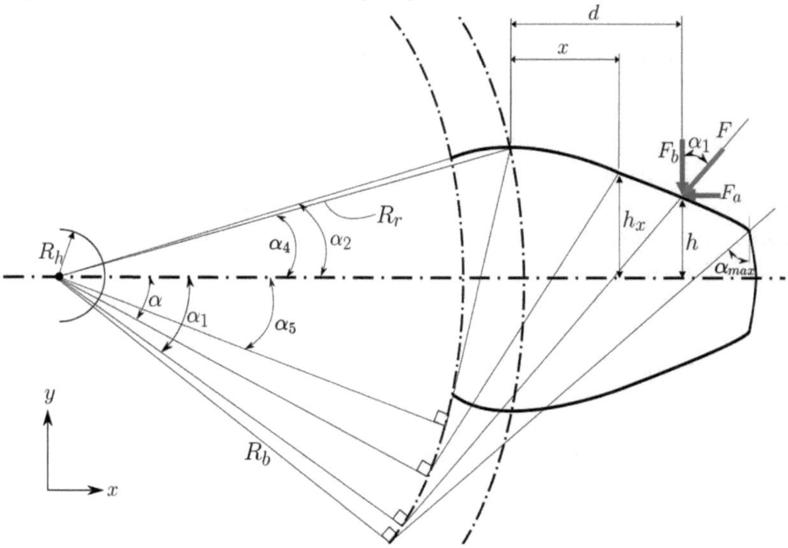

(b) Case 2: $R_r > R_b$. The angle between the intersection point of the root radius and the tooth gear and the centreline is given by α_4. Here α_{max} is the same as defined for case 1. Figure adapted from Van Eyk et al. (2022)

Fig. 3. Illustration of the two possible cases when modelling a gear. Case 1 represents the case where the base radius is larger than the root radius and case 2 the opposite.

This formulation assumes a constant height for any point before d_1. By substituting Eqs. (16)–(20) back into Eqs. (7)–(9) the following formulae are found:

$$\frac{1}{k_a} = \frac{2U_a}{F^2} = \int_0^{d_1} \frac{\sin^2\alpha_1}{2ELh_{x_1}} dx + \int_{-\alpha_1}^{\alpha_2} \frac{R_b(\alpha_2-\alpha)\cos\alpha\sin^2\alpha_1}{EA_x} d\alpha \tag{23}$$

$$= \frac{d_1\sin^2\alpha_1}{2ELh^*_{x_1}} + \frac{R_b\sin^2\alpha_1}{E}\int_{-\alpha_1}^{\alpha_2} \frac{(\alpha_2-\alpha)\cos\alpha}{A^*_x} d\alpha \tag{24}$$

$$\frac{1}{k_b} = \frac{2U_b}{F^2} = \int_0^{d_1} \frac{3\left[\cos\alpha_1(d-x) - \sin\alpha_1 h\right]^2}{2ELh^3_{x_1}} dx$$

$$+ \int_{-\alpha_1}^{\alpha_2} \frac{R_b^3\left\{1+\cos\alpha_1[(\alpha_2-\alpha)\sin\alpha - \cos\alpha]\right\}^2(\alpha_2-\alpha)\cos\alpha}{EI_x} d\alpha \tag{25}$$

$$= R_b^3 \frac{\left[1 - \frac{R_r}{R_b}\cos\alpha_1\cos\alpha_3\right]^3 - \left[1-\cos\alpha_1\cos\alpha_2\right]^3}{2EL\cos\alpha_1 h^{*3}_{x_1}}$$

$$+ \frac{R_b^3}{E}\int_{-\alpha_1}^{\alpha_2} \frac{\left\{1+\cos\alpha_1[(\alpha_2-\alpha)\sin\alpha - \cos\alpha]\right\}^2(\alpha_2-\alpha)\cos\alpha}{I^*_x} d\alpha \tag{26}$$

$$\frac{1}{k_s} = \frac{2U_s}{F^2} = \int_0^{d_1} \frac{2.4(1+\nu)\cos^2\alpha_1}{EA_x} dx + \int_{-\alpha_1}^{\alpha_2} \frac{2.4(1+\nu)R_b(\alpha_2-\alpha)\cos\alpha\cos^2\alpha_1}{EA_x} d\alpha \tag{27}$$

$$= \frac{1.2(1+\nu)d_1\cos^2\alpha_1}{ELh^*_{x_1}} + \frac{2.4(1+\nu)R_b\cos^2\alpha_1}{E}\int_{-\alpha_1}^{\alpha_2} \frac{(\alpha_2-\alpha)\cos\alpha}{A^*_x} d\alpha \tag{28}$$

where $*$ quantities refer to fault-adjusted quantities, which are described in the second part of this work. Although this set of equations is not novel in itself, the fact that h_x, A_x and I_x are not simplified, is novel. In the second part of this work, these quantities will be altered on a tooth-to-tooth basis based on the presence of faults.

Case 2: $R_r > R_b$: For case 2, two angular quantities are required, namely α_4 and α_5. To calculate α_5, the following set of equations (Liang et al. 2014) are simultaneously solved:

$$R_r\sin\alpha_4 = R_b\left[(\alpha_2-\alpha_5)\cos\alpha_5 - \sin\alpha_5\right] \tag{29}$$

$$R_r\cos\alpha_4 = R_b\left[(\alpha_5-\alpha_2)\sin\alpha_5 + \cos\alpha_5\right] \tag{30}$$

With some manipulation, rearranging and squaring, these two equations are summed to remove α_4 from the formulation, which leads to the following expression, which needs to be solved via Newton's method or any other suitable solver:

$$\left(\frac{R_b}{R_r}\right)^2\left[((\alpha_2-\alpha_5)\cos\alpha_5 - \sin\alpha_5)^2 + ((\alpha_5-\alpha_2)\sin\alpha_5+\cos\alpha_5)^2\right] = 1 \tag{31}$$

This equation has multiple stable roots, of which the root closest, but still smaller than α_2, is correct. From initial experimentation, it was found that initialisation of $\alpha_5 = 0$ always converged to the correct value for multiple different models. With some foresight, the value of α_4 will be required, and thus also needs calculation. This may be found by rearranging Eq. (29):

$$\alpha_4 = \arcsin\left(\frac{R_b}{R_r}((\alpha_2 - \alpha_5)\cos\alpha_5 - \sin\alpha_5)\right) \tag{32}$$

Transformation equations for case two can finally be determined. Equations (16), (21) and (22) remain unchanged and are therefore not repeated here:

$$d = R_b[(\alpha_1 + \alpha_2)\sin\alpha_1 + \cos\alpha_1] - R_r\cos\alpha_4 \tag{33}$$

$$x = R_b[(\alpha - \alpha_2)\sin\alpha + \cos\alpha] - R_r\cos\alpha_4 \tag{34}$$

$$h_x = R_b[(\alpha_2 - \alpha)\cos\alpha + \sin\alpha] \tag{35}$$

Once again, the stiffness integrals may be determined:

$$\frac{1}{k_a} = \frac{2U_a}{F^2} = \int_{-\alpha_1}^{\alpha_5} \frac{R_b(\alpha_2 - \alpha)\cos\alpha\sin^2\alpha_1}{EA_x} d\alpha \tag{36}$$

$$= \frac{R_b\sin^2\alpha_1}{E}\int_{-\alpha_1}^{\alpha_5} \frac{(\alpha_2 - \alpha)\cos\alpha}{A_x^*} d\alpha \tag{37}$$

$$\frac{1}{k_b} = \frac{2U_b}{F^2} = \int_{-\alpha_1}^{\alpha_5} \frac{R_b^3\left\{1 + \cos\alpha_1[(\alpha_2 - \alpha)\sin\alpha - \cos\alpha]\right\}^2(\alpha_2 - \alpha)\cos\alpha}{EI_x} d\alpha \tag{38}$$

$$= \frac{R_b^3}{E}\int_{-\alpha_1}^{\alpha_5} \frac{\left\{1 + \cos\alpha_1[(\alpha_2 - \alpha)\sin\alpha - \cos\alpha]\right\}^2(\alpha_2 - \alpha)\cos\alpha}{I_x^*} d\alpha \tag{39}$$

$$\frac{1}{k_s} = \frac{2U_s}{F^2} = \int_{-\alpha_1}^{\alpha_5} \frac{2.4(1 + \nu)R_b(\alpha_2 - \alpha)\cos\alpha\cos^2\alpha_1}{EA_x} d\alpha \tag{40}$$

$$= \frac{2.4(1 + \nu)R_b\cos^2\alpha_1}{E}\int_{-\alpha_1}^{\alpha_5} \frac{(\alpha_2 - \alpha)\cos\alpha}{A_x^*} d\alpha \tag{41}$$

Hertzian Stiffness (k_h): The Hertzian energy (U_h) is stored in the contact regions surrounding the line of contact between the two meshing teeth. This occurs due to the local deformation when two surfaces interact. The stiffness associated with the contact line between the two gear teeth is modelled with the contacting tooth faces assumed to be cylindrical (El Yousfi et al. 2019). The Hertzian energy term is defined as follows:

$$U_h = \frac{F^2}{2k_h} = \frac{F^2}{2} \times \frac{4(1 - \nu^2)}{\pi EL_x} \tag{42}$$

which leads to the equivalent stiffness term (Yang & Lin 1987):

$$\frac{1}{k_h} = \frac{2U_h}{F^2} = \frac{4(1-\nu^2)}{\pi E L_x^*} \tag{43}$$

Fillet Foundation Stiffness (k_f). The fillet foundation term accounts for the effect of the gear's body deflecting under a load on the tooth. This term is approximated by a parametric formula from Sainsot et al. (2004). The modelling procedure uses the notation defined in Fig. 4. The energy stored in tooth base deflection is based on a parametric equation from Sainsot et al. (2004):

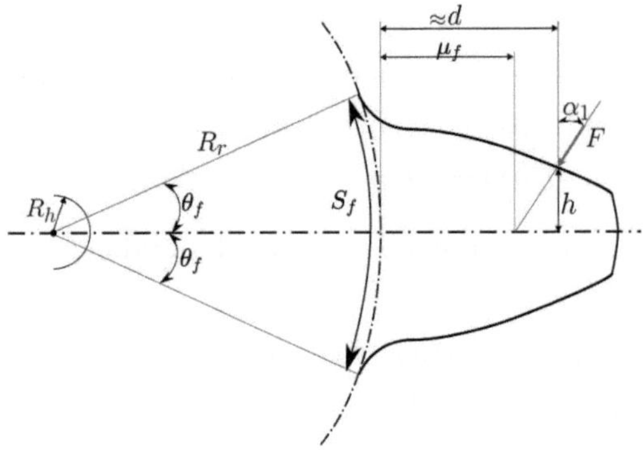

Fig. 4. Representation of the important parameters required to calculate the fillet foundation stiffness. Note that from Eqs. (17) and (33), d is defined relative to the point where the root circle and the tooth intersect. However, for the definition in Sainsot et al. (2004), the distance is measured relative to the intersection point of the tooth centreline with the root circle. Therefore, as a close approximation, d is used.

$$U_f = \frac{F^2}{2k_f} = \frac{F^2}{2} \times \frac{\cos^2\alpha_1}{EL_x}\left[L^*\left(\frac{\mu_f}{S_f}\right)^2 + M^*\left(\frac{\mu_f}{S_f}\right) + P^*(1+Q^*\tan^2\alpha_1)\right] \tag{44}$$

with the equivalent stiffness term given as:

$$\frac{1}{k_f} = \frac{2U_f}{F^2} = \frac{\cos^2\alpha_1}{EL_x}\left[L^*\left(\frac{\mu_f}{S_f}\right)^2 + M^*\left(\frac{\mu_f}{S_f}\right) + P^*(1+Q^*\tan^2\alpha_1)\right] \tag{45}$$

Note that faults do not modify the fillet foundation stiffness, and therefore, L_x is used instead of L_x^*. The μ_f and S_f parameters (according to Fig. 4) are defined as follows:

$$\mu_f = d - h \tan \alpha_1 \tag{46}$$

$$S_f = 2R_r\theta_f \tag{47}$$

$$\theta_f = \frac{1}{Z_k} \times \left[\frac{\pi}{2} + 2\tan \alpha_0 \times (h_a - \bar{r}_c) + \frac{2\bar{r}_c}{\cos \alpha_0}\right] \tag{48}$$

where $h_a = 1$ (as from the definitions in Eqs. (5) and (6)) and $\bar{r}_c = 0.2$ as taken from Sainsot et al. (2004) and is defined as the dimensionless tool tip radius.

Finally, L^*, M^*, P^* and Q^* are determined by substituting their coefficient values into the following equation, defined by Sainsot et al. (2004):

$$X^*(h_i, \theta_f) = \frac{A_i}{\theta_f^2} + B_i h_i^2 + \frac{C_i h_i}{\theta_f} + \frac{D_i}{\theta_f} + E_i h_i + F_i \tag{49}$$

where $h_i = R_r/R_h$. A_i, B_i, C_i, D_i, E_i and F_i may be found in Table 2.

Table 2. Coefficients required for the polynomial curve fit for fillet foundation stiffness (Sainsot et al.2004).

	A_i	B_i	C_i	D_i	E_i	F_i
L*(h, θ_f)	−5.574E−5	−1.9986E−3	−2.3015E−4	4.7702E−3	0.0271	6.8045
M*(h, θ_f)	60.111E−5	28.100E−3	−83.431E−4	−9.9256E−3	0.1624	0.9086
P*(h, θ_f)	−50.952E−5	185.50E−3	0.0538E−4	53.3000E−4	0.2895	0.9236
Q*(h, θ_f)	−6.204E−5	9.0889E−3	−4.0964E−4	7.8297E−3	−0.1472	0.6904

The fillet foundation stiffness assumes a single tooth pair is in contact. Many works add this term twice during double-tooth meshing, incorrectly treating each pair as independent, which leads to an overestimation of stiffness. To address this, Ma et al. (2016) proposed a FEM-derived calibration procedure. The FEM calibration procedure is an additional step that can be performed in conjunction with the proposed general stiffness model. However, the present focus is to establish a modelling framework rather than fine-tuning specific stiffness components, and therefore, the present work neglects the FEM calibration step.

2.3 Total Meshing Stiffness (k_t)

The total meshing stiffness for any pinion tooth at a given meshing cycle angle θ_i is calculated using (Chen & Shao 2011):

$$k_t(\theta_i) = \frac{1}{\frac{1}{k_h} + \frac{1}{k_{bm1}} + \frac{1}{k_{f1}} + \frac{1}{k_{bm2}} + \frac{1}{k_{f2}}} \tag{50}$$

where the subscripts 1 and 2 refer to the pinion and driven gear, respectively, and k_h is shared between the tooth pair. The beam stiffness term (k_{bmk}) combines axial, bending, and shear stiffness according to:

$$k_{bmk} = \frac{1}{\frac{1}{k_{a,k}} + \frac{1}{k_{b,k}} + \frac{1}{k_{s,k}}} \tag{51}$$

To calculate k_{bm1} and k_{f1} for the pinion, α_1 from Eq. (12) is used, while k_{bm2} and k_{f2} for the driven gear are calculated using α_1 from Eq. (13). Input parameters, such as R_b, must be adjusted to match the respective gear (e.g., $R_b = R_{b1}$ for the pinion and $R_b = R_{b2}$ for the driven gear). The meshing cycle angle θ_i always refers to the pinion, as in Eqs. (12) and (13).

2.4 Determining Meshing Stiffness for a Full Rotation of the Pinion Gear

The stiffness model is extended to a full pinion rotation ($\theta_p \in [0, 2\pi]$) by calculating the contribution of each tooth and combining the results.

Building Stiffness Arrays Independently for Each Tooth. For each tooth n, a stiffness array ($\mathbf{k_t}^{(n)}$) is constructed over $[0, 2\pi]$ by using:

$$k_t^{(n)}(\theta_p) = \begin{cases} k_t(\theta_{norm}), & \text{for } \theta_{norm} \in [0, \theta_t] \\ 0, & \text{otherwise} \end{cases} \tag{52}$$

where $k_t(\theta_i = \theta_{norm})$ is determined from Eq. (50) and θ_{norm} is defined by:

$$\theta_{norm} = \begin{cases} \theta_p + \theta_d + \theta_s, & \text{for } \theta_p \leq \theta_d \text{ and } n = Z_1 \\ \theta_p - \frac{2\pi}{Z_1}(n-1), & \text{otherwise} \end{cases} \tag{53}$$

This method creates zero vectors where stiffness values only replace these zeros during the tooth's meshing cycle. Special handling ensures the final tooth ($n = Z_1$) aligns correctly with the start of the stiffness array. This formulation simplifies practical implementation and avoids ambiguities.

Combining Tooth Stiffness Arrays: The individual stiffness vectors ($\mathbf{k_t}^{(n)}$) are superimposed to obtain the total meshing stiffness over a full pinion rotation:

$$\mathbf{k_t} = \sum_{n=1}^{Z_1} \mathbf{k_t}^{(n)} \tag{54}$$

Figure 5 illustrates the individual tooth stiffness vectors ($\mathbf{k_t}^{(n)}$) as dashed lines and the total stiffness ($\mathbf{k_t}$) as solid lines. Regions with overlapping contributions (double tooth meshing) result in higher stiffness, while single-tooth regions yield lower stiffness.

In summary, the current work has presented a simple yet effective framework for modelling gear mesh stiffnesses, with the further ability to alter these stiffnesses on a tooth by tooth basis with individual faults (demonstrated in the second part of this work). Herein lies the novelty of the current framework.

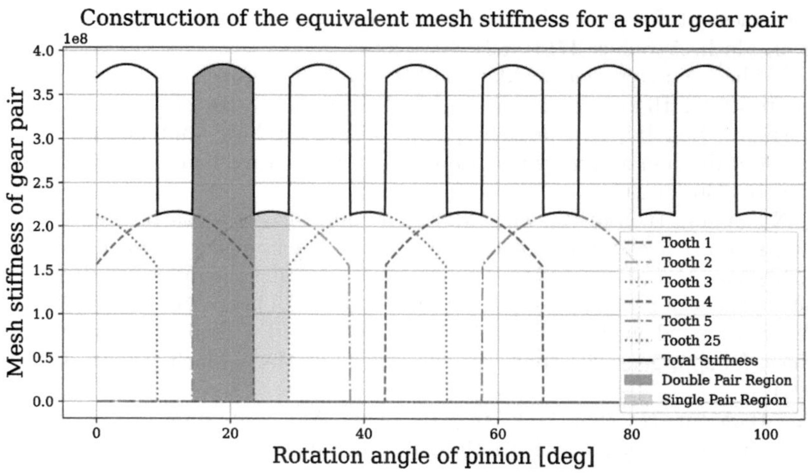

Fig. 5. Visualisation of individual stiffness arrays for each tooth ($\mathbf{k_t}^{(n)}$) along with the superimposed result ($\mathbf{k_t}$). To reconstruct this figure, utilise the parameters in Table 1.

3 Conclusions and Recommendations

This work introduces a novel generalised framework for gear mesh stiffness (GMS) modelling using the potential energy method (PEM). A key contribution is a simple yet effective stiffness combination formulation, which aggregates the contributions of all meshing teeth over a full rotation. This formulation overcomes ambiguities in previous works and provides an easily implementable method for GMS modelling.

This approach provides a robust foundation for further research and practical applications in gearbox modelling. The generalised framework simplifies the incorporation of fault models and is described in the second part of this work in a subsequent chapter.

The proposed framework has been converted to open-source Python code which can be accessed at: https://github.com/LukevanEyk/Gearbox-Simulator. This code allows users to input spur gear, fault and gearbox parameters and obtain GMS, fault visualisations and solved gearbox acceleration signals. The latter feature facilitates the generation of simulated gearbox data for machine learning.

Future work should address the calibration of the fillet foundation stiffness to enhance the model's precision. Additionally, extending the framework to planetary and helical gearboxes would broaden its applicability.

References

Chaari, F., Baccar, W., Abbes, M.S., Haddar, M.: Effect of spalling or tooth breakage on gearmesh stiffness and dynamic response of a one-stage spur gear transmission. Eur. J. Mech. A/Solids **27**(4), 691–705 (2008)

Chen, Z., Shao, Y.: Dynamic simulation of spur gear with tooth root crack propagating along tooth width and crack depth. Eng. Fail. Anal. **18**(8), 2149–2164 (2011)

El Yousfi, B., Soualhi, A., Medjaher, K., Guillet, F.: New approach for gear mesh stiffness evaluation of spur gears with surface defects. Eng. Fail. Anal. **116**(December 2019), 104740 (2020)

Liang, X., Zhang, H., Liu, L., Zuo, M.J.: The influence of tooth pitting on the mesh stiffness of a pair of external spur gears. Mech. Mach. Theory **106**, 1–15 (2016)

Liang, X., Zuo, M.J., Pandey, M.: Analytically evaluating the influence of crack on the mesh stiffness of a planetary gear set. Mech. Mach. Theory **76**, 20–38 (2014)

Liu, Y., Shi, Z., Shen, G., Zhen, D., Wang, F., Gu, F.: Evaluation model of mesh stiffness for spur gear with tooth tip chipping fault. Mech. Mach. Theory **158**, 104238 (2021)

Luo, Y., Baddour, N., Han, G., Jiang, F., Liang, M.: Evaluation of the time-varying mesh stiffness for gears with tooth spalls with curved-bottom features. Eng. Fail. Anal. **92**(June), 430–442 (2018)

Ma, H., Song, R., Pang, X., Wen, B.: Time-varying mesh stiffness calculation of cracked spur gears. Eng. Fail. Anal. **44**, 179–194 (2014)

Ma, H., Zeng, J., Feng, R., Pang, X., Wen, B.: An improved analytical method for mesh stiffness calculation of spur gears with tip relief. Mech. Mach. Theory **98**, 64–80 (2016)

Meng, Z., Wang, F., Shi, G.: A novel evolution model of pitting failure and effect on time-varying meshing stiffness of spur gears. Eng. Fail. Anal. **120**(February 2020), 105068 (2021)

Mohammed, O.D., Rantatalo, M.: Gear fault models and dynamics-based modelling for gear fault detection – a review. Eng. Fail. Anal. **117**(May), 104798 (2020)

Sainsot, P., Velex, P., Duverger, O.: Contribution of gear body to tooth deflections - a new bidimensional analytical formula. J. Mech. Des. Trans. ASME **126**(4), 748–752 (2004)

Sobie, C., Freitas, C., Nicolai, M.: Simulation-driven machine learning: bearing fault classification. Mech. Syst. Signal Process. **99**, 403–419 (2018)

Tian, X.: Dynamic simulation for system response of gearbox including localized gear faults. Masters dissertation, University of Alberta (2004)

Van Eyk, L.: A hybrid gearbox condition monitoring methodology using transfer learning for calibration. Technical report (2021)

van Eyk, L., Heyns, P.S., Schmidt, S.: A short review of gear fault modelling in a hybrid modelling context. In: Hammami, A., Heyns, P.S., Schmidt, S., Chaari, F., Abbes, M.S., Haddar, M. (eds.) MOSCOSSEE 2021. ACM, vol. 20, pp. 242–258. Springer, Cham (2022). https://doi.org/10.1007/978-3-030-85584-0_24

Wan, Z., Cao, H., Zi, Y., He, W., He, Z.: An improved time-varying mesh stiffness algorithm and dynamic modeling of gear-rotor system with tooth root crack. Eng. Fail. Anal. **42**, 157–177 (2014)

Yang, D.C., Lin, J.Y.: Hertzian damping, tooth friction and bending elasticity in gear impact dynamics. J. Mech. Des. Trans. ASME **109**(2), 189–196 (1987)

Yang, L., Wang, L., Yu, W., Shao, Y.: Investigation of tooth crack opening state on time varying meshing stiffness and dynamic response of spur gear pair. Eng. Fail. Anal. **121**(December 2020), 105181 (2021)

A Generalised Stiffness Model for Incorporating Multiple Spur Gear Tooth Faults - Part 2: Fault Modelling

Luke van Eyk(✉), P. Stephan Heyns, and Stephan Schmidt

Centre for Asset Integrity Management, University of Pretoria, Pretoria, South Africa
luke.vaneyk@up.ac.za

Abstract. Gear tooth faults significantly affect the meshing stiffness between gears and, consequently, the dynamic response of gearboxes. Accurate gear mesh stiffness (GMS) modelling is essential for studying tooth faults' influence in a condition-based maintenance context. This work presents the fault modelling portion of the generalised GMS framework described in the first of the two parts. Four distinct fault models—root cracks, chips, spalls, and pits—are implemented within the GMS framework, creating an environment where different fault types can be compared. The framework further provides a robust mechanism for generating labelled fault datasets, facilitating the development of hybrid data-driven and physics-based condition monitoring approaches. Fault models are integrated into the generalised stiffness model by modifying key stiffness parameters such as the contact length, effective tooth height, tooth's cross-sectional area and area moment of inertia. The model can introduce various fault types on individual teeth, with each tooth supporting only one fault type at a time. However, different faults can be applied to different teeth within the same gear, allowing for complex fault distributions that resemble real-world gearbox degradation. By consolidating fault modelling into a single structure, this work advances the capabilities of gearbox fault analysis and lays the foundation for future extensions and applications. This work is released with the corresponding open-source Python code at: https://github.com/LukevanEyk/Gearbox-Simulator.

Keywords: Generalised stiffness model · Potential energy method · Spur gear fault modelling · Time-varying mesh stiffness

1 Introduction

Gearboxes are widely used in industrial applications, including wind turbines, trucks, and helicopters (Chaari et al. 2008, Liu et al. 2021). Operating in harsh conditions, they are prone to failures, with gear tooth faults such as pitting, cracking, and spalling being the primary causes (Liang et al. 2016). These faults manifest as gear meshing stiffness (GMS) reductions, making their modelling

A. Hammami et al. (Eds.): MOSCOSSEE 2024, ACM 23, pp. 216–228, 2025.
https://doi.org/10.1007/978-3-032-00930-2_20

crucial for understanding gearbox dynamics and condition monitoring (Meng et al. 2021).

Gear tooth fault modelling has been extensively studied, with crack models (Tian 2004, Wu et al. 2008, Chen & Shao 2011), chip models (Chaari et al. 2008, Liu et al. 2021) and spall and pit models (Chaari et al.2008, Lei et al. 2018, El Yousfi et al. 2019) serving as representative models. Despite these advancements, the fault modelling literature is fragmented, with inconsistencies in how GMS reductions are incorporated for different fault types (van Eyk et al. 2022). Additionally, most works focus on single fault types, limiting their applicability in condition monitoring systems that aim to detect multiple fault types. A unified framework addressing these issues is therefore essential.

This work extends the generalised GMS framework from the first part of this work to incorporate fault models. The proposed framework supports the introduction of cracks, chips, spalls, and pits on individual gear teeth, with each tooth experiencing only one fault type at a time. However, multiple faults can be distributed across different teeth within the same gear, enabling complex fault scenarios. By integrating all fault models within a common framework, this approach allows for direct comparisons between different fault types and facilitates the generation of synthetic datasets for hybrid condition monitoring methods (Liu et al. 2021, Sobie et al. 2018).

The following section (Sect. 2) details the fault models and their integration into the GMS framework. The chapter then concludes with key findings and recommendations in Sect. 3.

2 Fault Modelling Within the Generalised Framework

2.1 Fault Introduction Methodology

Faults are introduced into the generalized stiffness model by modifying the L_x, h_x, A_x, and I_x parameters, which directly affect the stiffness terms in Eqs. (24), (26), (28), (37), (39), (41) and (43) from the first part of this work. For healthy teeth, these parameters retain their default values as defined by Eqs. (16)–(22) and (33)–(35) in the first part of this work. However, when a fault is present, these parameters are replaced by fault-specific values: $L_x^* = L_x^{fault}$, $h_x^* = h_x^{fault}$, $A_x^* = A_x^{fault}$, and $I_x^* = I_x^{fault}$, where the fault type determines the specific adjustments. This methodology explicitly incorporates fault-dependent terms into the stiffness equations, enabling the modelling of multiple fault types within the same framework. In this model, the driven gear remains fault-free, while faults can be introduced on any tooth of the pinion gear. Each tooth can have only one fault type, but different faults can be distributed across multiple teeth, enabling the simulation of diverse fault scenarios.

The stiffness equations require fault-adjusted expressions for L_x^*, A_x^*, I_x^*, and h_x^* as functions of x, which are transformed into angular functions via Eqs. (19) or (34) from the first part of this work. The following subsections describe four fault types, namely root cracks, chips, spalls, and surface pits, defining the necessary parameter adjustments for each case. Subsequent fault figures are constructed using the fault parameters from Table 1.

Table 1. Fault parameters used to generate the faults on the pinion gears in Figs. 3b, 5b, 7b, 9b, and 10.

Fault Type	Quantity Name	Symbol	Value	Unit
Crack	Left side crack depth	q_0	1.4	mm
	Right side crack depth	q_2	70	μm
	Crack angle	ν	90	°
	Crack width	L_c	N.A.	mm
Chip	Chip depth	b	3.5	mm
	Chip width	c	30.4	mm
Spall	Spall length	l_s	1.9	mm
	Spall width	w_s	30.4	mm
	Spall depth	h_s	2	mm
	Spall start location	x_1	6	mm
Pit	Sampled pit location	u_i	$\mathcal{N}(4.36,\, 5\text{E}{-3})$	mm
	Sampled pit contact angle	ϕ_i	$\mathcal{N}(\frac{\pi}{4}, \frac{\pi}{60})$	°
	Sampled pit radius	r_i	$\mathcal{N}(0.5, 0.15)$	μm
	Number of samples	N.A.	50	-

2.2 Crack Modelling

The crack model utilised in this work is based on Chen & Shao (2011). In their approach, cracks are modelled as parabolic profiles ($q(z)$) representing depth variations across the tooth width and split into two scenarios: partial-width cracks (solid line) and full-width cracks (dashed line), as shown in Fig. 1a. For partial-width cracks:

$$q_1(z) = \begin{cases} q_0\sqrt{\frac{L_c - z}{L_c}}, & z \in [0, L_c] \\ 0, & z \in [L_c, L] \end{cases} \tag{1}$$

and for full-width cracks:

$$q_2(z) = \sqrt{\frac{q_2^2 - q_0^2}{L} z + q_0^2} \tag{2}$$

where z describes the position on the tooth width, $q_i(z)$ defines the variable crack depth for case 1 or 2 and L_c represents the crack end width on the full tooth width, L. Crack depth at the left and right side of the gear tooth are indicated by the q_0 and q_2 parameters (illustrated in Fig. 1c), respectively.

The crack depth q_j is normalised using the crack plane angle ν from Fig. 1b:

$$q_j = q_{j,plane} \cdot \sin(\nu) \tag{3}$$

where q_j indicates the normalised slice depth (see q_j in Fig. 1c). The actual quantity of interest, namely the effective tooth height, h_q may be determined from:

$$h_q = h_c - q_j, \quad h_c = R_b \sin(\alpha_2) \tag{4}$$

(Chen & Shao 2011) solve the stiffness integrals for each width slice j and sum the results:

$$k_t = \sum_{j=1}^{J} k_t^j \tag{5}$$

where k_t refers to the total meshing stiffness from equation (50) in the first part of this work.

To simplify this approach, a novel equivalent analytical average effective crack height, $h_{q,ave,k}$, may be used. The effective crack height is calculated using:

$$h_{q,ave,k} = \frac{1}{L} \int_0^L h_q(z)dz \tag{6}$$

where k refers to the crack scenario, yielding:

$$h_{q,ave,1} = h_c - \left(\frac{2L_c q_0}{3L}\right) \sin(\nu) \tag{7}$$

$$h_{q,ave,2} = h_c - \left(\frac{2}{3} \times \frac{q_2^3 - q_0^3}{(q_2^2 - q_0^2) + \epsilon}\right) \sin(\nu) \tag{8}$$

where ϵ prevents division by zero for $q_0 = q_2$. The equivalent crack height for both scenarios is illustrated in Fig. 2. The resulting fault-adjusted parameters for stiffness calculations become:

$$h_x^{crack} = \begin{cases} h_x, & h_x \leq h_{q,ave,k}, \\ \max\left(0, \frac{h_x + h_{q,ave,k}}{2}\right), & h_x > h_{q,ave,k} \end{cases} \tag{9}$$

$$A_x^{crack} = 2L(h_x^{crack}), \quad I_x^{crack} = \frac{2L}{3}(h_x^{crack})^3 \tag{10}$$

Figures 3a and 3b demonstrate the application of this average model for a full-width crack and its impact on GMS. The crack model can also simulate a broken tooth by setting $q_0 = q_2 = 2h_c$.

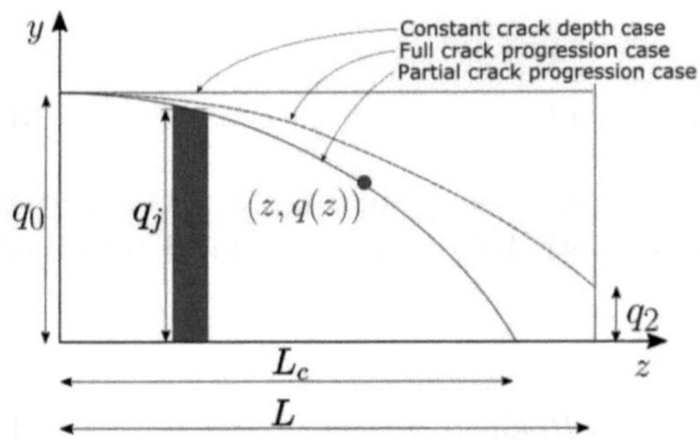

(a) Crack width model adapted from van Eyk et al. (2022).

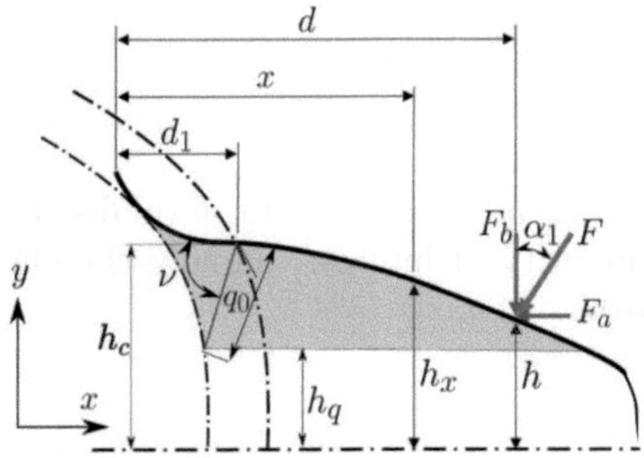

(b) Effective tooth height due to crack stiffness reduction adapted from van Eyk et al. (2022).

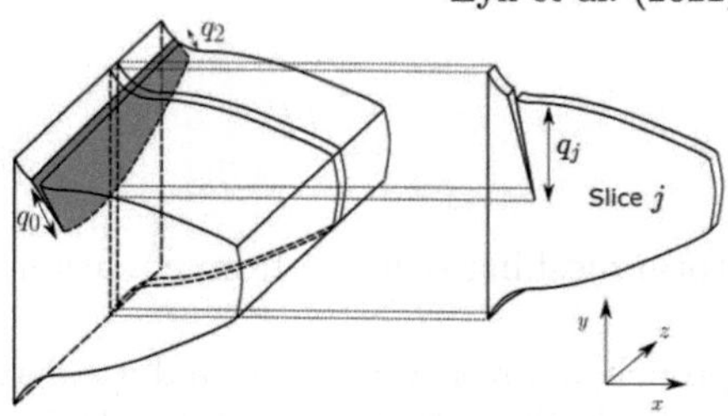

(c) Illustration of the three-dimensional nature of the root crack. Note that the crack is induced with varying width from one edge (q_0) to the other edge (q_2). On the right-hand side of the figure is a single slice j, which results in a uniform crack depth of q_j.

Fig. 1. Illustration of (a) the formal notation of the methodology from Chen & Shao (2011), (b) the traditional model used to define constant width cracks and (c) a three-dimensional view of the methodology proposed by Chen & Shao (2011) to improve upon the traditional crack modelling method.

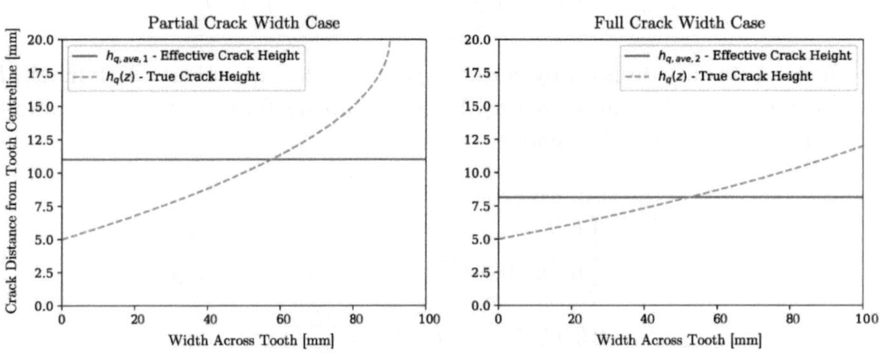

Fig. 2. Demonstration of the new crack formulation method where a varying crack height profile is converted into its average counterpart.

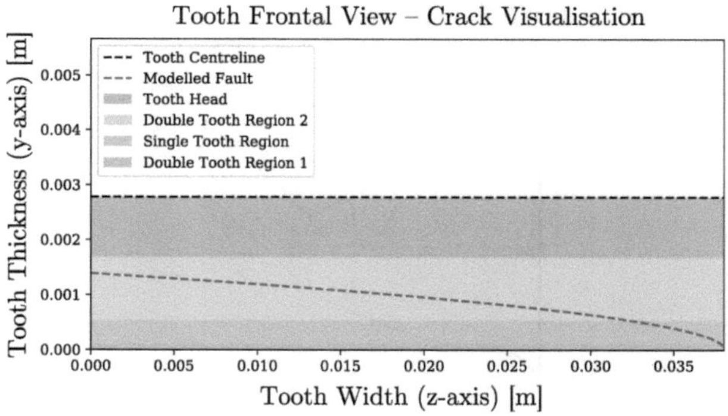

(a) A visualisation of a crack induced on a pinion gear tooth (Frontal view of tooth).

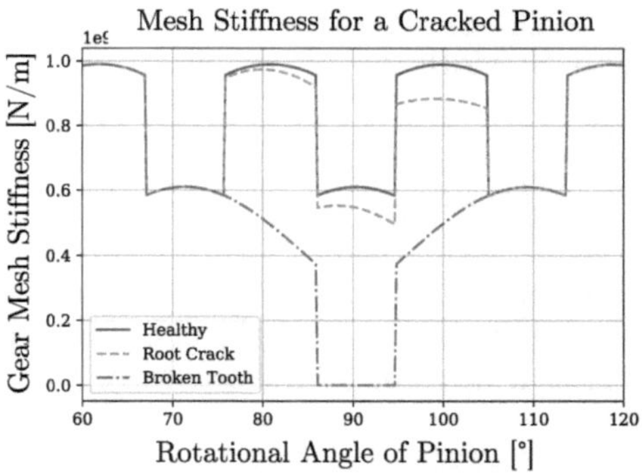

(b) Resulting GMS from a cracked and broken pinion gear tooth.

Fig. 3. Effect of crack or broken tooth on gear meshing stiffness.

2.3 Chip Modelling

This work demonstrates the chip model proposed by Tian (2004), which models the chip profile as a hyperbola and disregards depth effects. This approach simplifies the analysis as the chip fault only affects the tooth width (L_x^*) and, consequently, the Hertzian stiffness term. The chip profile is characterised by the chip depth along the tooth face (b) and the width across the tooth tip (c), as illustrated in Fig. 4. The chip formulation requires defining the maximum tooth length, d_h, measured from the root and is dependent on the root (R_r) and base (R_b) radii:

$$d_h = \begin{cases} R_b[(\alpha_{max} - \alpha_2)\sin\alpha_{max} + \cos\alpha_{max}] - R_r\cos\alpha_3, & R_r \leq R_b \\ R_b[(\alpha_{max} - \alpha_2)\sin\alpha_{max} + \cos\alpha_{max}] - R_r\cos\alpha_4, & R_b < R_r \end{cases} \tag{11}$$

where α_{max} is the angular position corresponding to the maximum x value, calculated as $\alpha_{max} = \alpha_1(\theta_{tot})$ from Eq. (12) in the first part of this work. Using

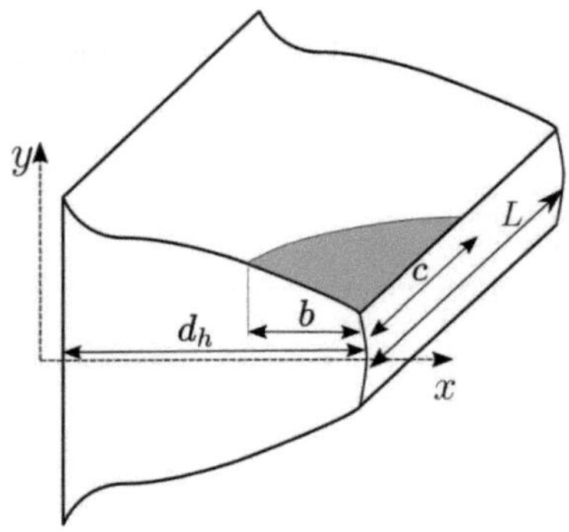

Fig. 4. Chip model utilised in this work. Adapted from Tian (2004).

d_h, the chip-adjusted tooth width is:

$$L_x^{chip} = \begin{cases} L_x, & 0 \leq x \leq d_h - b \\ L_x - \left(\frac{d_h c}{b} - \frac{1}{x}\left[\frac{d_h^2 c}{b} - d_h c\right]\right), & d_h - b \leq x \leq d_h \end{cases} \tag{12}$$

For a given angle α, the corresponding x value determines which case is applied in Eq. (12). Figures 5a and 5b demonstrate the effect of a chip on a pinion gear tooth and the resulting gear meshing stiffness (GMS).

2.4 Spall Modelling

The spall model used in this work is based on Chaari et al. (2008), which represents the spall as a rectangular region with constant depth. This model accounts for changes in the tooth width (L_x^*), cross-sectional area (A_x^*), and area moment of inertia (I_x^*), affecting both the Hertzian and beam stiffness terms. The spall is characterised by its width (w_s), length (l_s), depth (h_s), and starting location (x_1), as shown in Fig. 6.

The spall formulation is defined as:

$$L_x^{spall} = \begin{cases} L_x, & x \in [0, x_1] \text{ or } [x_1 + l_s, d_h] \\ L_x - w_s, & x \in [x_1, x_1 + l_s] \end{cases} \tag{13}$$

$$A_x^{spall} = \begin{cases} A_x, & x \in [0, x_1] \text{ or } [x_1 + l_s, d_h] \\ A_x - h_s w_s, & x \in [x_1, x_1 + l_s] \end{cases} \tag{14}$$

$$I_x^{spall} = \begin{cases} I_x, & x \in [0, x_1] \text{ or } [x_1 + l_s, d_h] \\ I_x - \frac{w_s}{12} h_s^3, & x \in [x_1, x_1 + l_s] \end{cases} \tag{15}$$

Here, $0 \leq x_1 \leq d_h$, where d_h is defined in Eq. (11). This formulation requires four parameters—l_s, w_s, h_s, and x_1—to fully define the spall's size and location.

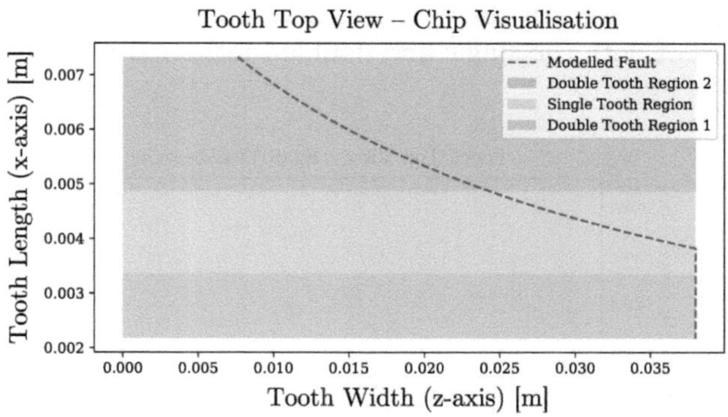

(a) Visualization of a chip induced on a pinion gear tooth (Top view).

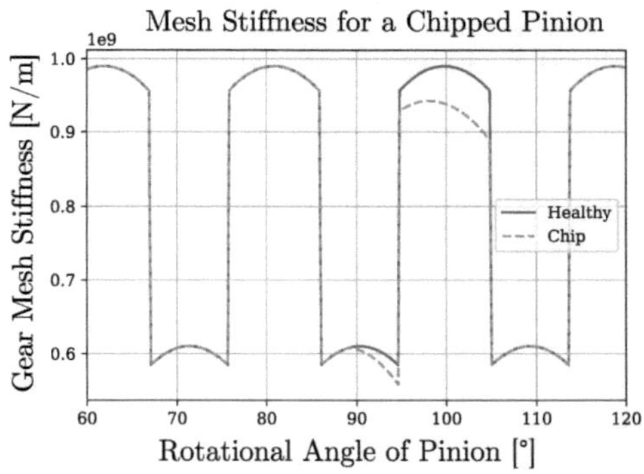

(b) Resulting GMS from a chipped pinion gear tooth.

Fig. 5. Effect of a chip on a pinion gear tooth and the resulting GMS.

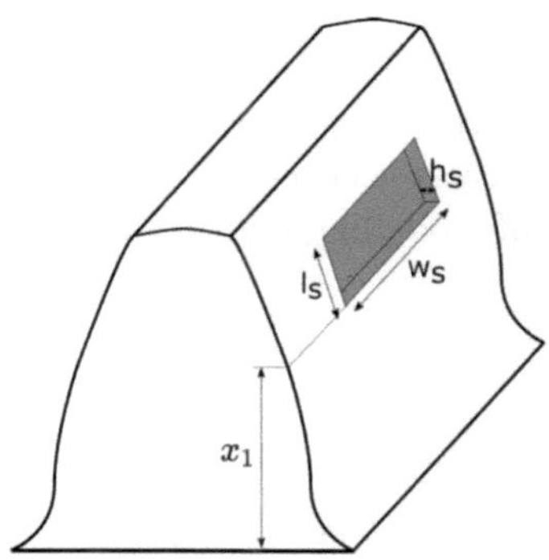

Fig. 6. Spall model utilised in this work. Based on Chaari et al. (2008) and adapted from van Eyk et al. (2022).

Figures 7a and 7b demonstrate the impact of an induced spall on the tooth face plane and its effect on the gear meshing stiffness (GMS). The spall fault affects portions of both the single and double-tooth meshing regions.

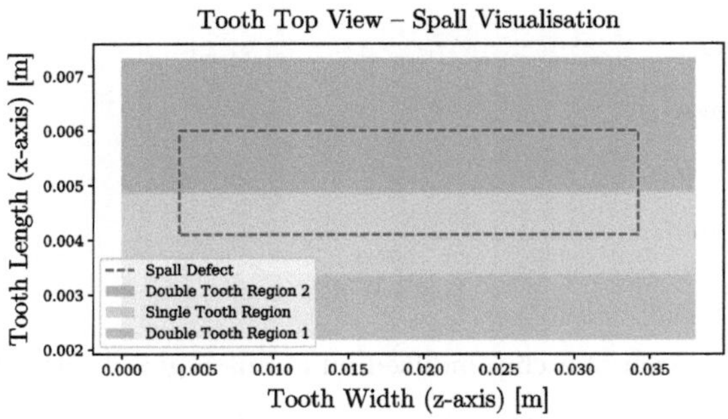

(a) Visualization of a spall induced on a pinion gear tooth.

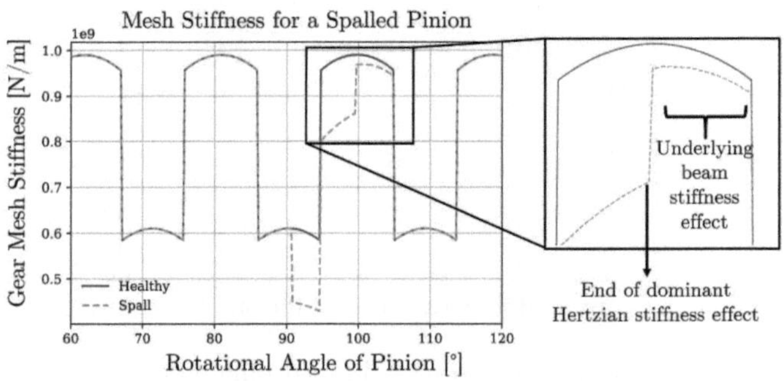

(b) Resulting GMS from a spalled pinion gear tooth.

Fig. 7. Effect of a spall on a pinion gear tooth and its resulting GMS.

Figure 7b reveals a reduction in GMS from 91° to 100° due to the Hertzian stiffness term. Beyond 100°, the reduction in bending and axial stiffness persists due to the reduced effective cross-sectional area and area moment of inertia. This highlights the importance of modelling multiple stiffness effects for accurate fault representation.

2.5 Pit Modelling

The pitting model used in this work is based on Meng et al. (2021), representing pits as spheres that subtract material from the gear tooth. Each pit is characterised by a centre location (u), radius (r), and contact angle (ϕ), as shown in Fig. 8a. The contact angle defines how far the sphere intersects the tooth surface, allowing for flexible pit geometries ranging from shallow, wide pits (small ϕ, large r) to deep, narrow pits (large ϕ, small r).

Pits are generally clustered in the single tooth meshing region due to higher stresses in this area. To model these clusters, the pit centres (u_i) are sampled from a normal distribution:

$$u_i \sim \mathcal{N}(u_p, \sigma_u) \tag{16}$$

where u_p and σ_u represent the mean and spread of the distribution. Pit sizes are also sampled from normal distributions for the radius (r_i) and contact angle (ϕ_i):

$$r_i \sim \mathcal{N}(r_p, \sigma_r) \tag{17}$$

$$\phi_i \sim \mathcal{N}(\phi_p, \sigma_\phi) \tag{18}$$

where r_p and ϕ_p define the mean values, and σ_r and σ_ϕ their standard deviations.

Figure 8b visualises an ensemble of sampled pits. The variation in location is shown by the normally distributed faults, whilst the shape variation is shown as different-sized and coloured pits across the tooth width.

Each pit's effect on the tooth is calculated individually as reductions in contact length (ΔL_x), cross-sectional area (ΔA_x), and area moment of inertia (ΔI_x):

$$\Delta L_{x,i} = \begin{cases} 2\sqrt{(r_i \sin\theta_i)^2 - (u_i - x)^2}, & x \in [u_i - r_i \sin\theta_i, u_i + r_i \sin\theta_i] \\ 0, & \text{otherwise} \end{cases} \tag{19}$$

$$\Delta R_i = \begin{cases} \sqrt{r_i^2 - (u_i - x_i)^2}, & x \in [u_i - r_i \sin\theta_i, u_i + r_i \sin\theta_i] \\ 0, & \text{otherwise} \end{cases} \tag{20}$$

$$\Delta A_{x,i} = \frac{\Delta R_i^2 (2\theta_i - \sin 2\theta_i)}{2} \tag{21}$$

$$\Delta I_{x,i} = \frac{\Delta R_i^4}{72}\left(18\theta_i - 9\sin 2\theta_i \cos 2\theta_i - \frac{64\sin^6\theta_i}{2\theta_i - \sin 2\theta_i}\right) + \frac{A_x \Delta A_{x,i}\left(h_x - \left(\frac{4\Delta R_i \sin^3\theta_i}{3(2\theta_i - \sin 2\theta_i)} - \Delta R_i \cos\theta_i\right)\right)^2}{(A_x - \Delta A_{x,i})} \tag{22}$$

Deltas are summed over all N pits to calculate the total fault-adjusted values:

$$L_x^{pit} = L - \sum_{i=1}^{N} \Delta L_{x,i} \tag{23}$$

$$A_x^{pit} = A_x - \sum_{i=1}^{N} \Delta A_{x,i} \tag{24}$$

$$I_x^{pit} = I_x - \sum_{i=1}^{N} \Delta I_{x,i} \tag{25}$$

Figures 9a and 9b illustrate a cluster of pits and their impact on the GMS. Larger, older pits appear in lighter colours, while smaller, newer pits are darker.

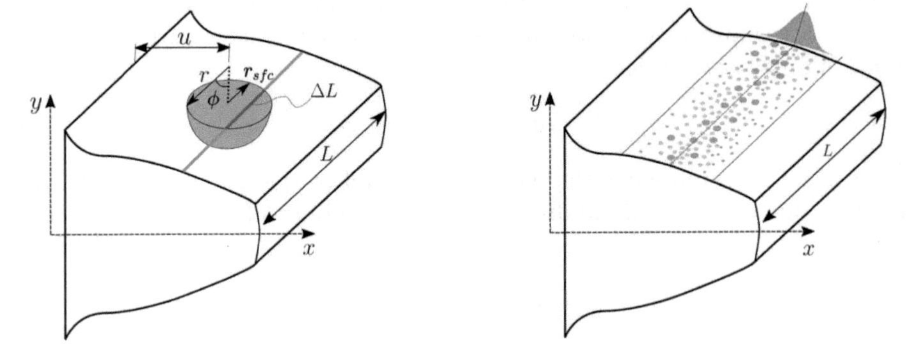

(a) Pit model utilized in this work. Size is exaggerated for clarity.

(b) Pit distribution model followed in this work.

Fig. 8. Individual and distributed pit models based on Meng et al. (2021).

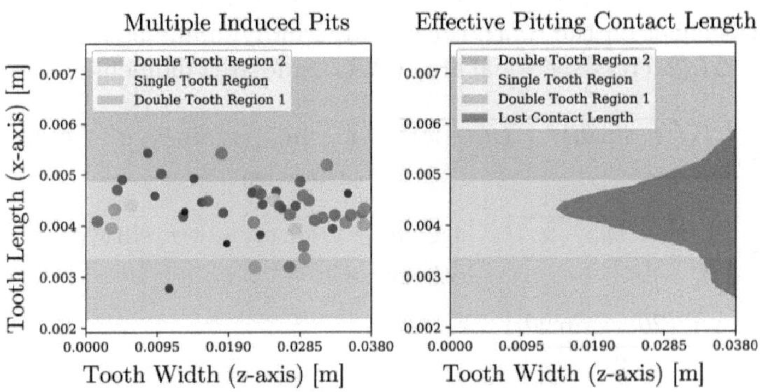

(a) Visualization of pits induced on a pinion gear tooth.

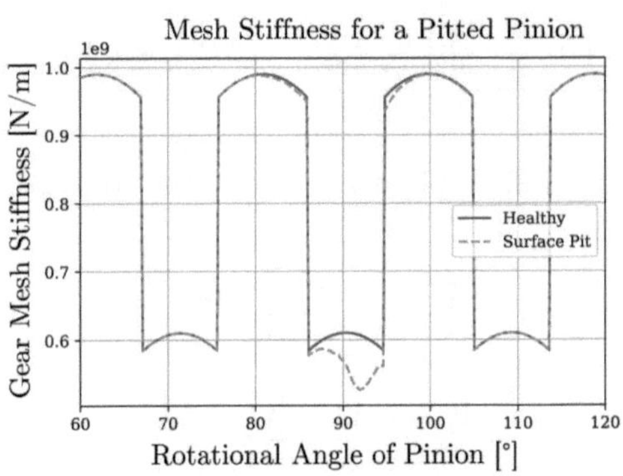

(b) Resulting GMS from a pitted pinion gear tooth.

Fig. 9. Effect of pits on a pinion gear tooth and the resulting GMS.

2.6 Illustrative Example of Multiple Gear Tooth Faults

To demonstrate the flexibility of the proposed method, five different faults are introduced on the same pinion gear, spaced two teeth apart, as shown in Fig. 10. Note that the broken tooth fault is an extreme case of the crack model and not a distinct fault type. This figure highlights the novelty of the generalised model, which allows for the simultaneous introduction of multiple fault types on different gear teeth.

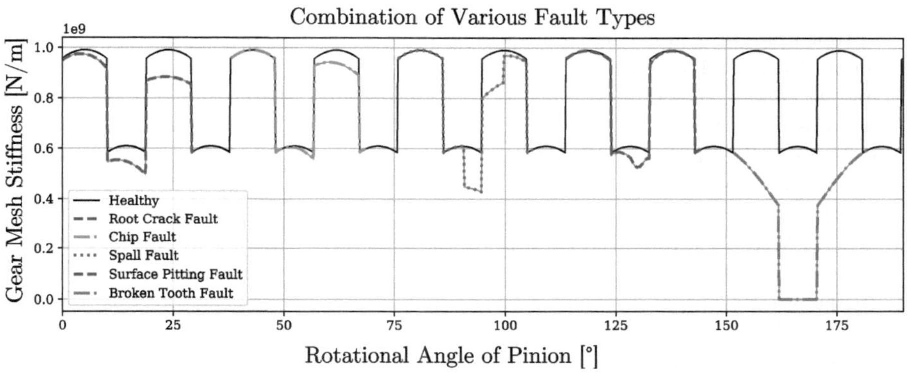

Fig. 10. Multiple fault models implemented on the same pinion gear, showcasing the flexibility of the proposed generalised model to simulate diverse fault types. Fault parameters are listed in Table 1.

3 Conclusions and Recommendations

This work focuses on the fault modelling aspect of the generalised stiffness framework introduced in Part 1. The model expands existing approaches by enabling the simulation of multiple fault types, each applied to different teeth within the same spur gear. This work demonstrates four distinct fault models—root cracks, chips, spalls, and pits—within the proposed generalised stiffness modelling structure, allowing for direct comparison of faults and flexible fault placement.

The framework further serves as a powerful mechanism to generate labelled fault datasets, which can support the development of hybrid-based condition monitoring approaches. To further amplify the impact of this work, an open-source Python repository is made available so that researchers can reproduce or extend the results from the proposed generalised framework and accompanying faults. The code is hosted at: https://github.com/LukevanEyk/Gearbox-Simulator.

Future work should focus on updating existing fault models to incorporate state-of-the-art models. Furthermore, other fault models not considered here can be developed according to the framework.

References

Chaari, F., Baccar, W., Abbes, M.S., Haddar, M.: Effect of spalling or tooth breakage on gearmesh stiffness and dynamic response of a one-stage spur gear transmission. Eur. J. Mech. A/Solids **27**(4), 691–705 (2008)

Chen, Z., Shao, Y.: Dynamic simulation of spur gear with tooth root crack propagating along tooth width and crack depth. Eng. Fail. Anal. **18**(8), 2149–2164 (2011)

El Yousfi, B., Soualhi, A., Medjaher, K., Guillet, F.: New approach for gear mesh stiffness evaluation of spur gears with surface defects. Eng. Fail. Anal. **116**(December 2019), 104740 (2020)

Lei, Y., Liu, Z., Wang, D., Yang, X., Liu, H., Lin, J.: A probability distribution model of tooth pits for evaluating time-varying mesh stiffness of pitting gears. Mech. Syst. Signal Process. **106**, 355–366 (2018)

Liang, X., Zhang, H., Liu, L., Zuo, M.J.: The influence of tooth pitting on the mesh stiffness of a pair of external spur gears. Mech. Mach. Theory **106**, 1–15 (2016)

Liu, Y., Shi, Z., Shen, G., Zhen, D., Wang, F., Gu, F.: Evaluation model of mesh stiffness for spur gear with tooth tip chipping fault. Mech. Mach. Theory **158**, 104238 (2021)

Meng, Z., Wang, F., Shi, G.: A novel evolution model of pitting failure and effect on time -varying meshing stiffness of spur gears. Eng. Fail. Anal. **120**(February 2020), 105068 (2021)

Sobie, C., Freitas, C., Nicolai, M.: Simulation-driven machine learning: bearing fault classification. Mech. Syst. Signal Process. **99**, 403–419 (2018)

Tian, X.: Dynamic simulation for system response of gearbox including localized gear faults. Masters dissertation, University of Alberta (2004)

van Eyk, L., Heyns, P.S., Schmidt, S.: A short review of gear fault modelling in a hybrid modelling context. In: Hammami, A., Heyns, P.S., Schmidt, S., Chaari, F., Abbes, M.S., Haddar, M. (eds.) MOSCOSSEE 2021. ACM, vol. 20, pp. 242–258. Springer, Cham (2022).

Wu, S., Zuo, M.J., Parey, A.: Simulation of spur gear dynamics and estimation of fault growth. J. Sound Vib. **317**(3–5), 608–624 (2008)

Author Index

A
Abdelmoula, Zied 36
Akrout, Ali 137, 190
Amor, Walid Ouled 13

B
Baccar, Abderrahim 137, 190
Benali, Salwa 168
Bouaziz, Slim 90, 168

C
Chaari, Fakher 118, 126, 137, 190
Cheniour, Houssem 90
Chikhaoui, Chaima 154

D
Daoud, Hajer 146
Dhieb, Youssef 13
Djemal, Fathi 154, 178
Dourado, Nuno 137, 190

E
ElMahi, Abderrahim 146

G
Ghariani, Moez 13

H
Haddar, Maroua 3, 26, 98
Haddar, Mohamed 3, 36, 54, 80, 90, 98, 126, 146
Hammami, Ahmed 3, 26, 36, 54, 80, 98, 118, 126, 137, 190
Hattay, Amira 54
Hentati, Taissir 168
Heyns, P. Stephan 200, 216
Hmida, Fatma 178

I
Ibarbia, Alberto Diez 80

M
Mahmood, Yaser S. H. 64
Mahmoud, Abdel Kader 54, 98, 126
Maktouf, Rania 26
Mbarek, Ayoub 118, 126
Meddeb, Firas 146
Mohameden, Cheikh Brahim 98

O
Omar, Rawnak 154, 178

R
Rebiere, Jean-Luc 146
Rincon, Alfonso Fernandez Del 80
Rueda, Fernando Viadero 80

S
Samet, Ahmed 109
Schmidt, Stephan 200, 216
Souf, Mohamed Amine Ben 146

T
Taktak, Mohamed 64
Toumi, Abdesselem 80
Trabelsi, Hassen 154, 178

V
van Eyk, Luke 200, 216

Y
Yahya, Ahmed Mohamed 54, 98, 126
Yangui, Majdi 90

A. Hammami et al. (Eds.): MOSCOSSEE 2024, ACM 23, p. 229, 2025.
https://doi.org/10.1007/978-3-032-00930-2

MIX
Papier aus verantwortungsvollen Quellen
Paper from responsible sources
FSC® C105338

If you have any concerns about our products,
you can contact us on
ProductSafety@springernature.com

In case Publisher is established outside the EU,
the EU authorized representative is:
Springer Nature Customer Service Center GmbH
Europaplatz 3, 69115 Heidelberg, Germany

Printed by Libri Plureos GmbH
in Hamburg, Germany